ENVIRONMENTAL IMPACT ASSESSMENT

A Guide to Best Professional Practices

ENVIRONMENTAL IMPACT ASSESSMENT

A Guide to Best Professional Practices

CHARLES H. ECCLESTON

CRC Press
Taylor & Francis Group
Boca Raton   London   New York

CRC Press is an imprint of the
Taylor & Francis Group, an **informa** business

CRC Press
Taylor & Francis Group
6000 Broken Sound Parkway NW, Suite 300
Boca Raton, FL 33487-2742

First issued in paperback 2017

ISBN 13: 978-1-4398-2873-1 (hbk)
ISBN 13: 978-1-138-07415-6 (pbk)

This book contains information obtained from authentic and highly regarded sources. Reasonable efforts have been made to publish reliable data and information, but the author and publisher cannot assume responsibility for the validity of all materials or the consequences of their use. The authors and publishers have attempted to trace the copyright holders of all material reproduced in this publication and apologize to copyright holders if permission to publish in this form has not been obtained. If any copyright material has not been acknowledged please write and let us know so we may rectify in any future reprint.

Library of Congress Cataloging-in-Publication Data

Eccleston, Charles H.
    Environmental impact assessment : a guide to best professional practices / Charles Eccleston.
        p. cm.
    Includes bibliographical references and index.
    ISBN 978-1-4398-2873-1 (hardcover : alk. paper)
    1. Environmental impact analysis. 2. Environmental impact analysis--United States. I. Title.

TD194.6.E255 2011
333.71'4--dc22   2010037122

**Visit the Taylor & Francis Web site at
http://www.taylorandfrancis.com**

**and the CRC Press Web site at
http://www.crcpress.com**

About the cover

The photograph on the cover of this book was taken of deep space from the Hubble orbiting telescope. It has been said that in all the vastness of space, Earth is the only planet known to be capable of supporting advanced life. Perhaps nothing captures this concept and the delicate balance of our ecosystem more than this sole picture. This little speck of a planet is all we have. Like the inhabitants of Easter Island who were condemned to suffer the ravages of their reckless resource depletion, Planet Earth is also a finite and fragile ecosystem. Easter Islanders devastated their tropical paradise and eventually reaped the ultimate punishment for doing so. On a global scale, our actions could lead to a similar and ominous fate. The Earth is but a speck in an incomprehensible vastness of space, and like the harsh lesson of Easter Island, we will ultimately have to live with the consequences of our actions. Like the Easter Islanders, if we mismanage our resources, we have no other planet to flee to. It is with this thought in mind, that this book has been written. This book provides best professional practices (BPP) for performing impact assessments that will ultimately inform policy and decision makers about the consequences of potentially destructive actions before a final decision has been made. Ultimately, we will "reap what we sow." Let us sow with foresight and prudence.

Dedication

This book is dedicated to Alice and Brandie, who have been my inspiration.

Contents

Foreword

Charles Eccelston has established himself as one of the top environmental practitioners and teachers in the world. In Environmental Impact Assessment: A Guide to Best Professional Practices, Eccleston takes our profession's leading challenges, such as cumulative effects, greenhouse emissions, and environmental justice, and gives us drill-down best practice methodologies that are applicable to all U.S. and international assessment work. In an ever growing green culture and economy, the author's canon of work has global significance for accurately predicting project development outcomes and ways to rethink our infrastructure needs and reality. As an added bonus, Eccleston, who has been on the cutting edge of thought and practice related to ISO 14001 and environmental management systems (EMS), shows us how to use EMS for precedence-setting decision making. When an author can draw forth the importance of spirituality to social impact assessment and the relationship of black swans to accident probability analysis, you know you are in the presence of a master.

Ron Deverman
President, National Association of
Environmental Professionals
Collingswood, New Jersey, USA

Acknowledgments

I am indebted to the professional practitioners who reviewed and provided comments on this book. Although space constraints make it difficult to mention all individuals by name, I would like to call attention to the following professionals.

I am indebted to Peyton Doub, a seasoned NEPA practitioner and close associate, who reviewed Chapter 1 ("Cumulative Impact Assessment: A Synopsis of Guidance and Best Professional Practices") and Chapter 6 ("Environmental Management Systems").

I would like to express my thanks to Grace Musumeci of the US Environmental Protection Agency, who reviewed and provided insightful comments on Chapter 2 ("Preparing Greenhouse Emission Assessments") and Chapter 3 ("Preparing Risk Assessments and Accident Analyses").

Charles (Chuck) Nicholson reviewed and made important contributions to Chapter 2 ("Preparing Greenhouse Emission Assessments"), Chapter 4 ("Social Impact Assessment and Environmental Justice"), and Chapter 5 ("The International Environmental Impact Assessment Process").

I am also indebted to Judith (Judy) Charles who reviewed and provided comments on Chapter 1 ("Cumulative Impact Assessment: A Synopsis of Guidance and Best Professional Practices") and Chapter 2 ("Preparing Greenhouse Emission Assessments").

Introduction

Under the best of circumstances, environmental impact assessment (EIA) can be a complex and challenging task. Experience indicates that the scope and quality of such analyses varies widely throughout the U.S. as well as internationally. This book integrates five distinct yet interrelated themes into a single comprehensive framework for practitioners:

- Cumulative impact assessment
- Preparing greenhouse emission assessments
- Preparing risk assessments and accident analyses
- Social impact assessment and environmental justice
- The international environmental impact assessment process guiding principles

This book also describes the ISO 14001 environmental management system (EMS) and explains how it can be used to implement decisions that result from the aforementioned assessments; direction is provided for integrating the EMS with an international EIA process and the goal of sustainability. The thrust of the book is to provide practitioners and decision makers with *best professional practices* (BPP) for preparing such analyses. The aim of this book is to provide the reader with a balanced skill set of concepts, principles, and practices for these assessments. This book is unique in that it focuses on providing practitioners and decision makers with state-of-the-art tools, techniques, and approaches for resolving environmental impact assessment problems.

While the book references the U.S. National Environmental Policy Act (NEPA), most of this guidance is generally applicable to any international EIA process consistent with NEPA. A sixth and final chapter provides direction for developing a comprehensive Environmental Management Systems which can be used to monitor and implement final decisions based on such analyses.

Black Swans

Nassim Taleb developed the theory of Black Swan events. Taleb developed the theory to explain: 1) potentially rare but catastrophic, and difficult to predict events that lie beyond normal expectations and (2) the psychological biases that tend to blind people to the possibility of such uncertain events. A good example of a Black Swan event was the Deepwater Horizon drilling rig blowout. It was indeed difficult to predict every decision or event that led to one of the worst environmental disasters in modern history. Many critics charge, "What in the world were they thinking!" In hindsight, it is easy to claim that this disaster was predictable. In reality, it was not for this was a Black Swan event because it was extremely rare and not easily predictable with any degree of reasonable certainty.

The sequence of events that create routine environmental problems tend to be quite predictable, and are, therefore, termed White Swans. While most environmental White Swan events do not attract international attention, they can nevertheless cost millions of dollars in damage and fines, lead to loss of life, and ruin local ecosystems, to say nothing of ruining careers. Environmental catastrophes can still occur because there may be a near total absence of information that defines the ramifications of specific substances or operating practices that later turn out to be very harmful to the environment.

In the Deepwater Horizon, much of the BP blowout leak was simply due to a limited understanding of the limitations of shutoff equipment mounted one mile below the surface of the ocean. However, White Swan events involving more mundane or routine environmental issues are typically the result of a lack of awareness, inattentiveness, sloppiness, or trying to shortcut the safety/environmental process to save time or money. In such cases, a single or select few individuals are viewed as the "environmental people" and employee training is limited to the absolute basic elements; senior managers may feel unable to step in for fear of suffering serious career ramifications; the focus is frequently on complying with minimal environmental regulations rather than carefully planning out and considering all potential concerns. When an accident occurs, the innocent may be fired or demoted to demonstrate that swift action was taken to prevent a future event.

Both Black and White Swan events often have the underlying theme of a lack of cohesive leadership either just before the event or in the wake of the resulting crisis when everyone is panicked and responding to the event. Environmental departments routinely deal with a broad and cross-cutting array of departments. Environmental managers are often perfectly positioned to see the warning signs and to assume the leadership necessary to prevent such an event. Most importantly, they are in a unique position which exceeds the simpler task of day-to-day environmental auditing, inspections, and environmental compliance. They are in a position to perform detailed analyses of potential scenarios and their impacts, and to develop plans and mitigation measures for dealing with them should they occur. It is partly for this reason, that that this book has been written. This book is designed to provide environmental managers and analysts alike with assessment tools necessary for assessing and developing plans that will prevent not only White Swans, but Black Swan events as well.

The Legal Circle

To a large extent, the modern environmental movement was driven by public anger which led to enactment of strict laws and regulations and, inevitably, litigation.* Lawyers that helped to lead this movement were an idealistic breed. In their defense, corporations started hiring lawyers to address regulatory compliance issues and potential liability. This new generation of lawyers has become part of the business establishment. These corporate lawyers tend to view environmental compliance in terms of promoting the interests of the organizations that hire them. Virtually every company claims to be pursuing the goal of sustainable development future, while at the same time employing armies of lawyers to protect their interests.

Corporate managers and staff are being cautioned to carefully review memos and e-mails that may have even a remote possibility of being "discovered" as part of a lawsuit. Environmental, health, and safety (EHS) managers attend training classes which teach them how to think like lawyers. But this can also result in negative implications. While a manager may take the position of "remaining silent," minimizing important communications — this can also result in negative effects such as failure to communicate potential risks.

Many lawyers view their role as investigating every conceivable legal avenue to represent their clients and to minimize risks to the client. Juries are left with the complicated process of trying to sort out the facts and reach a conclusion. Attorneys are masters at manipulating juries and acquitting guilty defendants who then go on to commit even more heinous crimes. These lawyers are skilled at exploiting legal loopholes. One of the revelations that came out of the BP Deepwater Horizon oils spill was that those in charge of making decisions and oversight had not equipped the rig with a second, backup device intended to cut off the flow of oil from a well in case the blowout preventer failed.

* This article was inspired by an article by Richard MacLean, *Environmental Quality Management*, 117-123, Autumn 2010.

While such redundancy is common on new drilling rigs, it was not required under U.S. law so BP claims that they were in compliance with U.S. requirements. Unfortunately, this fact does not matter to the public. As a result, many attorneys may advise their clients to adopt the most expensive and conservative technologies and practices all the time, in all instances. But this is not necessarily a desirable outcome either. Managers may simply err on the most conservative side of every decision, dramatically reducing future business ventures. History has show time and again that some of the most successful development projects were the result of corporations taking risks and bold actions.

An attorney-dominated organization can be at peril when attorneys are granted too much control. The decision-making process can become corrupted where an organization's attorneys act to block access to upper management in an attempt to shield leaders from liability. In addition to listening to their attorneys, managers must also consider common industrial practices, and consider what the moral and ethically right thing to do is.

Organizational ethics are frequently interpreted within the narrow confines of existing regulations. This is particularly true of an area like sustainability where opinions vary widely. Green marketing has become very popular of late. Commitments are typically steeped in dazzling terms such as future benefits. One is left to wonder how these core principles are truly integrated into day-to-day operations. Organizational lawyers are often playing an integral part in such marketing.

This brings us back to the lawyers that had much to do with initiating the modern environmental movement. While some lawyers led this movement, many now specialize in circumventing environmental health, safety, and environmental quality; they counsel managers on how protect their organizations while wreaking havoc on the environment. So we have come almost full circle. As one lawyer commented, the best business lawyers think like business mangers and thus are not risk averse.

However, there is another avenue available to government and business for reducing risks; this approach can optimize decision-making while circumventing many of the chaotic and paradoxical legal dilemmas just described. This avenue involves preparing scientifically-based assessments which objectively evaluate decision making in terms of potential impacts, risks, and reasonable alternatives to what may be a standard or traditional course of action. Properly prepared, such assessments can provide managers and decision makers with a powerful tool for balancing the risks and impacts against more traditional factors such as cost and schedules. It is with this thought in mind that this book has been written.

Audience

This book is designed for use by practitioners and decision-makers who are faced with the challenge of preparing complex EIAs. The book is also aimed at professionals in government and consulting, and those in the private sector who are involved in some way with preparing NEPA or EIAs, and who seek to master this subject. If you have technical questions or issues, or need assistance, you may contact the author at Eccleston@ msn.com.

chapter one

Cumulative impact assessment: A synopsis of guidance and best professional practices

Not all chemicals are bad. Without chemicals such as hydrogen and oxygen, for example, there would be no way to make water, a vital ingredient in beer.

—Dave Barry

The US Council on Environmental Quality (CEQ) has indicated that there is increasing evidence that the most destructive environmental effects may actually result not from the direct and indirect effects of a given action, but instead from the combination of individual minor effects of numerous actions over time.[1] The CEQ's cumulative effects handbook recognizes the "cumulative effects analysis as an integral part of the National Environmental Policy Act (NEPA) process, not a separate effort."[2] Cumulative impact assessment (CIA) is as necessary and as much of a challenge in environmental assessments (EAs) as it is in environmental impact statements (EISs). In fact, CIA may be an even greater challenge in EAs, which are usually prepared for relatively small actions whose cumulative impacts are not always as evident as they are for the larger projects analyzed in EISs. Moreover, the issue of cumulative impacts has become one of the most widely litigated issues under the NEPA. The high number of challenges over cumulative impacts is likely to continue, if not increase, well into the future.

While, the CEQ handbook, *Considering Cumulative Effects under the National Environmental Policy Act*, provides guidance for addressing cumulative impacts, this chapter provides a compendium of best professional practices (BPPs) for complying with the CEQ's direction.[3] More to the point, CEQ's guidance focuses on "what" should be done, while this chapter centers on "how" a CIA should be prepared. One of the principal goals of this chapter is to provide practical and defensible direction that can reduce the future rate of this litigation. A detailed consideration of CIA is therefore appropriate and necessary in a book directed primarily at preparation of EAs. However, the text below provides a comprehensive discussion of CIA with direction that is applicable to both EAs and EISs. Although this chapter focuses on preparing NEPA CIAs, most of its content is also equally applicable to most international environmental impact assessments (EIAs).

As in the case of EISs, the analysis of cumulative effects provides a powerful tool in EAs for taking into account incremental impacts on environmental resources as federal officials plan future actions. Cumulative impacts are rarely as immediately evident as are direct, and even indirect, impacts attributable to single actions. As in EISs, agencies have become adept at analyzing direct and indirect impacts, but cumulative impacts have posed much more difficult scoping, analytical, and methodological problems, which have led to a host of legal challenges. The CEQ adds that "federal agencies routinely address the direct effects (and to a lesser extent indirect effects) of the proposed action on the environment. Analyzing cumulative effects is more challenging, primarily because of the difficulty of defining geographic (spatial) and time (temporal) boundaries for the analysis."[4]

Definition of cumulative impact

The CEQ NEPA regulations define cumulative impact as:

> *...the impact on the environment that results from the incremental impact of the action when added to other past, present, and reasonably foreseeable future actions, regardless of which agency (federal or non-federal) or person undertakes such other actions. Cumulative impacts can result from individually minor but collectively significant actions taking place over a period of time*

In other words, CIA must consider an action's incremental (i.e., direct and indirect) impacts combined with the effects of other past, present, and reasonably foreseeable future actions, to provide the decision maker and the public with a fuller understanding of the overall significance and consequence that can be expected in the future. By mandating an analysis of cumulative impacts, the regulations ensure that the range of actions considered in EAs and EISs account for not only the project proposal, but also other actions that could cumulatively harm environmental quality. The results of the CIA should be incorporated into the agency's overall environmental planning process. To this end, federal agencies are to use results obtained from CIA as a tool for evaluating the implications of a proposal in even *project-specific* EAs.[5]

Note that cumulative impacts are not the same as indirect impacts. The CEQ NEPA regulations define indirect impacts as:

> *...caused by the action and are later in time or farther removed in distance, but are still reasonably foreseeable.*

Indirect effects may include growth inducing effects and other effects related to induced changes in the pattern of land use, population density or growth rate, and related effects on air and water and other natural systems, including ecosystems.

Indirect impacts, like direct impacts, are "caused by the action." Cumulative impacts are not limited to those caused by the action under consideration. They are not even limited to impacts attributable to federal actions—nor are they limited to actions occurring at the same time as the action under consideration. They encompass the totality of impact to the affected resource, including those directly or indirectly attributable to the federal action under consideration (the proposed action), other federal actions, actions by state and local governments, and actions by private entities—whether occurring in the past, concurrently, or in the future. Consider the following two examples and the direct, indirect, and cumulative impacts associated with each:

Example 1: Proposed Action: Construction of a highway
- Example of a direct impact: Natural habitat losses caused by grading the roadbed
- Example of an indirect impact: Habitat losses caused by commercial development at a new interchange
- Example of a cumulative impact: Habitat losses caused by various development activities within a region

Example 2: Proposed Action: Allocation of federal water to a new power plant developed by investor-owned utility
- Example of a direct impact: Water contaminate increase in lake receiving cooling water discharges
- Example of an indirect impact: Water contaminate increase due to runoff from parking lot for power plant
- Example of a cumulative impact: Water contaminate increase from runoff from future lakeside housing projects built to house power plant employees

Other cumulative impact definitions

The following discussion uses the US NEPA definition when discussing the cumulative effects, unless otherwise indicated. In more straightforward terms, cumulative effects are defined as *the changes to the environment caused by an activity in combination with other past, present, and reasonably foreseeable human activities.* As used below, the terms "cumulative impact" and "cumulative effect" are used synonymously.

However, other definitions are in common use in other countries. Despite some differing terminology, most resemble the US NEPA definition in general concept. In 1988, the Canadian Environmental Assessment Research Council (CEARC) defined cumulative effects as those that occur when impacts on the natural and social environments take place so frequently in time or so densely in space that the effects of individual projects cannot be assimilated. They can also occur when the impacts of one activity combine with those of another in a synergistic form. The Canadian Environmental Assessment Act indicates that the EIA process should include consideration of:

> *...any cumulative environmental effects that are likely to result from the project in combination with other projects or activities that have been or will be carried out, and the significance of the effects.*

The following definition is routinely applied in much of Europe:

> *Cumulative impacts refer to the accumulation of human-induced changes in valued environmental or ecosystem components (VEC) across space and over time; such impacts can occur in an additive or interactive manner.*

Regardless of the precise definition employed, the concept of cumulative impacts derives from the observation that an impact of a particular project on an environmental resource may be considered insignificant when assessed in isolation; yet, the total or cumulative impact may be quite significant when evaluated in the context of the combined effect of all past, present, and reasonably foreseeable future activities that may have or have had an impact on the resources in question. For a more in-depth discussion of the analytical and regulatory requirements, the reader is referred to the companion text entitled *Environmental Impact Statements*.[6]

Types of cumulative impacts

Cumulative impacts can generally be divided into two broad classes: additive and synergistic (or interactive).

Additive cumulative impacts

Additive effects occur when the magnitude of combined effects is equal to the sum of individual effects. Common examples of additive effects encountered in CIA for NEPA EAs and EISs include:

- Multiple air emission sources affecting regional air quality
- Multiple point and non-point discharges to a watershed

Table 1.1 Daily Use of Water from a Reservoir for a proposed Cogeneration Facility.

Facility	Water withdrawn (mgd)	Water returned (mgd)	Net water use (mgd)	Reservoir storage required (acre-ft.)	Percent of total reservoir storage
Existing water users					
Burlington textile plant	6.30	5.70	0.60	108	0.01
Kerr Lake Regional Water System	6.00	5.40	0.60	108	0.01
Town of Clarksville	0.50	0.45	0.05	9	<0.01
Mecklenburg Correctional Facility	0.06	0.00	0.06	23	<0.01
City of Virginia Beach	60.00	0.00	60.00	10,200	0.99
Subtotal	**72.86**	**11.55**	**61.31**	**10,448**	**1.01**
Proposed water users					
Mecklenburg Cogeneration Plant (proposed action)	3.14	0.75	2.39	430	0.04
Virginia Power	25.20	2.10	23.10	4,158	0.40
Subtotal	**28.34**	**2.85**	**25.49**	**4,588**	**0.45**
Total Kerr Reservoir use	**101.20**	**14.40**	**86.80**	**15,036**	**1.46**

- Multiple water withdrawals from a river basin or aquifer
- Multiple losses of forest cover (or of other habitats) in a landscape

CIA for additive effects is conceptually simple. The assessor need only sum the magnitudes of effect and then assess the combined effect as if it resulted from a single project. For example, an EA for a coal-fired power plant summed the proposed water withdrawals for that project and for other federal, state, and private water withdrawals from the reservoir (Table 1.1).[7] It was possible to model the combined effect on water levels on the available storage capacity of the reservoir.

As another example, some analyses have treated salt deposition from multiple power plant cooling towers as additive. Cooling towers are used at many power plants to reduce the temperature of water used for steam generation prior to discharge to natural water bodies. Salts carried in plumes of vapor emitted from the cooling towers can damage vegetation if their concentrations exceed certain threshold levels. Consider an action to construct a new power plant and cooling towers on the outer edges of a property containing an existing power plant and cooling towers. The obvious approach to CIA when considering cumulative regional salt deposition is to model the combined deposition of the existing plus proposed new cooling towers.

Assessing some additive effects can, however, be considerably more difficult. For example, while a quantitative assessment of the summed effect of point source water discharges from multiple industrial operations to a surface water body may be possible, quantification of non-point source discharges such as runoff from farms and lawns is difficult and sometimes impossible. In such a case, assessors must sometimes fall back into qualitative analyses. The qualitative assessment might cite numerical data on projected site development from a comprehensive plan or land use plan, but conversion of projected site numbers into resulting nitrogen or phosphate discharges into a watershed is difficult. Any attempt at quantification would inevitably be forced to rely on broad worst-case assumptions.

Even more difficult is assessing additive effects that are inherently nonquantifiable. For example, visual impacts are usually impossible to assess quantitatively. Even if state-of-the-art digital models are used to project a future visual image of a proposed new facility, the image can usually only be presented with some descriptive text. Multiple new projects within the same viewshed (area visible from a reference point) result in additive cumulative impacts. Logically, many new facilities (of similar general size and character) constructed in the same viewshed result in greater cumulative visual impacts than only a few new facilities. But if direct and indirect impacts cannot be distilled to numbers, then the additive impact cannot be expressed as the sum of such numbers. The CIA, just like the evaluation of direct and indirect impacts, can only be descriptive. The need to consider the combined impacts is, however, no less important.

Synergistic cumulative impacts

When effects are combined, the result may be substantially greater or less than that expected based on additivity. A greater than expected result can be described as synergistic, and a less than expected result can be described as antagonistic. Because synergism is of greater concern than antagonism in assessing environmental impacts, a useful approach to CIA is to consider only additive and synergistic effects. Synergistic effects are usually much more complex and difficult to assess than additive effects. Often they cannot be expressed quantitatively. They generally result from interactions of two or more activities that result in combined effects that are greater than the sum of the individual project effects. For example, the release of two different chemicals into a stream may cause an interactive effect between the chemicals that is greater than what the individual effects would be.

As an example, consider that the combined effects of terrestrial habitat losses can be expected to be additive when the habitat type is abundant in the region, but can be synergistic when the habitat type is regionally scarce. For example, bird species favoring large expanses of forest cover become increasingly scarce once landscapes lack forest tracts below area thresholds that differ by species. Incremental forest losses in a well-forested landscape might be considered to constitute an additive loss of forest habitat, but the incremental losses that reduce the largest remaining forest tracts to sizes below the area thresholds can have a substantially greater effect on forest interior bird species than accounted for by the forest area losses alone.

Additive or synergistic, cumulative impacts that can alter environmental systems through multiple pathways of effect include the following:[8]

- *Growth-inducing pathway.* As described above, each new project can induce further actions or projects to occur, sometimes referred to as "spin-off" effects (e.g., an off-trail road resulting in increased hunting and fishing). The best example of the growth-inducing pathway is the potential for urban development along a new highway project. Highways reduce commuting times between rural areas and urban business centers, thereby making residential development more financially attractive. Areas around interchanges ("cloverleaves") and intersections become especially attractive to commercial development. Such induced development is a common consideration in NEPA documents for highway projects. The fact that the residential and commercial development is not federal is not germane to a CIA for a federally financed or permitted highway project. Low levels of growth-induced cumulative impacts tend to be additive while higher levels have a greater potential to be synergistic.
- *Physical or chemical transport pathway.* A physical or chemical constituent is transported away from the activity under review, where it then interacts with another activity (e.g., air emissions, sedimentation, wastewater effluent). Multiple point and non-point discharges of water pollutants are typical examples. Multiple air emission sources are also common examples. As for other pathways, cumulative impacts tend to be additive at lower levels and synergistic at higher levels. However, the interaction of multiple chemicals in the environment can be complex, and synergistic impacts could result at even low levels of total impact.
- *Nibbling loss pathway:* This is the gradual disturbance and loss of land or habitat (e.g., clearing of land for a new subdivision and

new roads into a forested area). The presence of an action nearly always leads to some degree of landscape disruption, representing a "nibbling" loss of land potential to support other uses; it is this type of cumulative effect that cannot always be easily addressed on a project-by-project review basis. Numerous small actions within the same area can cause cumulative (nibbling) effects. Unlike many large development projects that can be effectively master planned to minimize environmental impacts, the numerous small projects are often uncoordinated and haphazard, with none of the participating developers ever considering the overall impacts from the totality of activity. Often this happens when many developments occur in rapid succession (e.g., development boom). These types of actions may cause far more cumulative effects than a single coordinated action in the same area. The nibbling loss pathway is not always sharply delineated from the growth-induced pathway. The situation of multiple land development projects induced by the opening of a new highway. Growth-induced pathway can also be interpreted as an example of the nibbling loss pathway. The additive water use impacts provided as an example of additive cumulative impacts in Table 1.1 can be interpreted as following the nibbling loss pathway, but where water rather than land or natural habitat is the nibbled resource.

Normally, the impacts of nibbling cannot be adequately dealt with in terms of an individual project-review basis. While regional changes can often be quantified (e.g., fragmentation of wildlife habitat or total cleared land), it is more difficult to determine a significance to this change that is only attributable to the specific action under review. To properly address this type of cumulative effect, regional plans are required that clearly establish regional thresholds of change against which the specific actions may be compared. Many counties and other municipalities develop comprehensive plans that propose an overall desired direction for future development that considers the totality of land use conflicts and environmental impacts.

The scale problem: defining spatial and temporal boundaries

One of the principal reasons that CIA tends to be so much more challenging than the corresponding assessment of direct and indirect effects is simply the difficulty of defining the geographic (spatial) and time (temporal) boundaries. If these boundaries are defined too broadly, the analysis becomes exhausting, unwieldy, and perhaps excessively alarming. Conversely, if they are defined too narrowly, the analysis may be insufficient to inform decision makers of potentially significant cumulative impacts.

Impact assessments have traditionally involved defining more or less arbitrary boundaries around action sites that are often local and limited to the effects of the single action. CIA, by definition, expands those traditional spatial and temporal horizons. The appropriate spatial and temporal scale for each CIA depends on the specific issue being addressed. No single scale is generally appropriate for all resources or issues, or even for distinct types of impacts on the same resource. Many CIAs have been deemed deficient because either the spatial scale was too small or the temporal scale too short, or both. The implication of too small a boundary is that important spatial or long-term effects may be neglected. For example, the effects of a housing project on regional traffic patterns might be minimized if a CIA considers only development within a small incorporated city and not the surrounding county or planning area. The implication of too large a boundary is that the localized importance of effects may be excessively minimized when viewed against an excessively broad area (essentially a dilution of effects). For instance, the traffic increases from the housing project and other projects in the surrounding county might appear to be trivial when considered in the context of traffic issues faced by an entire state. Often, CIAs are arbitrarily limited to political or jurisdictional boundaries, or individual ownerships, or an arbitrarily chosen time frame such as 10 years. Relying on jurisdictional borders to define the study area may be expedient but such an approach generally ignores the ecological realities. For example, to determine boundaries for assessing water quality, one may "trace" the path of a chemical constituent along a river as far as one believes it may still be reactive, causing a significant effect.

The Supreme Court has emphasized that agencies may properly limit the scope of their CIA based on practical considerations. Here, the court wrote:

> *Even if environmental interrelationships could be shown conclusively to extend across basins and drainage areas, practical considerations of feasibility might well necessitate restricting the scope of comprehensive statements.*[9]

Importance of proper scoping

Scoping plays a key role in focusing and defining the analysis of cumulative impacts, just as for direct and indirect impacts. Scoping can be used to identify spatial boundaries, time frames, and key issues or effects to consider. A well-orchestrated scoping process provides the best opportunity to identify important cumulative impact assessment issues, setting appropriate boundaries for analysis, and identifying relevant past, present, and future actions for investigation. Scoping also facilitates interagency cooperation needed to identify agency plans and other actions whose effects may overlap with those of the impacts of the proposed action.

Fortunately, determining the appropriate spatial bounds is somewhat more straightforward than the selection of the temporal (timeframe) bounds. For example, when considering cumulative impacts on a rare plant or animal species, spatial analysis scale might be defined to encompass the current or historical distribution range of that species. For the example of cumulative water withdrawals from a reservoir, the reservoir and its associated watershed (land area contributing precipitation to the reservoir and its tributaries) might be the defined as the spatial bound. The situation is usually more complex: There may be hundreds of rare species to consider, and the appropriate scales may be different for each species.

As noted above, the appropriate hydrologic units for assessing urban runoff are generally the watershed and sub-watershed. The watershed is generally the largest confined hydrologic management unit. The sub-watershed comprises the land draining to the point where two second-order streams join, generally an area between 1 and 10 square miles in humid temperate zones such as the eastern United States. Care must be exercised to ensure that CIA considers the combined effects of activities in individual sub-watersheds that might be experiencing greater pollution or runoff than the watershed at large. However, the CIA must not ignore the combined effects of pollution and runoff from an individual sub-watershed to the downstream receiving waterways.

As another example, consider a project in which construction effects relating to noise, dust, and soil disturbance should dissipate to background levels within 1 mile. However, visual effects during operation would extend as much as 5 miles, depending on topography, viewing location, and weather influences such as fog. The CIA might, therefore, use a spatial boundary of 1 mile when analyzing noise, dust, and soil impacts, but a scale of 5 miles for the analysis of visual aesthetic impacts.

In general, the spatial boundary of a cumulative effects assessment should consider the limit, if any, to which a significant effect can reasonably be evaluated. Practitioners must determine at what point an effect becomes trivial or nonsignificant. Experienced analysts often establish boundaries based on the *zone of influence* (or *region of influence*), beyond which the effects of the action have dissipated to levels of a trivial or insignificant state. Ideally, such an approach should be taken for each effect on each environmental resource examined (e.g., water, air, wildlife, vegetation), therefore requiring multiple boundaries instead of the more typical single study area. Boundaries therefore expand and contract according to the unique issues and resources.

The concept that such a point is reached at a certain threshold is appealing, but often difficult to define in practice. The complexity of any relationship beyond those purely at the physical or chemical level often results in considerable reliance on best professional judgment and the consideration of risk. An iterative approach may need to be applied when setting boundaries, in which the first boundary, often arrived at by a professional "guess," may later change if new information suggests that a different boundary is warranted.

Arguments have been raised in some cases that the boundary should be national, or perhaps even international. However, such a scale is rarely merited and would usually be appropriate only for air or water effects (e.g., the long-range transport of air pollutants or greenhouse gases) or where species migrate considerable distances. In particular, evaluations of impacts on migratory birds in CIA may have to consider trends in both the summer "breeding" grounds, wintering grounds, and connecting flyways. For many species termed "neotropical migrants," the breeding grounds are in North America but the wintering grounds are in Central or South America.

On a more practical basis, boundaries may be assigned based on the limits of available data. A well-studied watershed or available coverage of remote sensed imagery may influence the spatial extent of an assessment because the cost and time required to obtain more data may be prohibitive and may not be justified by the scope of the final decision. The decision as to whether additional data is necessary requires the practitioners to judge the adequacy of existing data in providing the basis for a defensible decision.

Similarly, the appropriate time scales can also vary. There is no hard and fast rule governing how far into the future a CIA should be carried out. However, the further back or ahead in time, the greater the dependence will be on qualitative analysis and conclusions due to lack of descriptive information (e.g., what conditions were like years ago or which other actions may occur in the future) and increasing uncertainty in predictions. Thus, in practice, the scenario in the past often defaults to the year in which the baseline information for the assessment was collected (i.e., current conditions) and the future extends no further than including known (i.e., certain) actions.

In practice, the choice of the assessment time frame for future cumulative impacts is a matter of professional judgment. However, all assumptions used in defining the temporal and spatial boundaries for a CIA must be clearly stated and justified.

For instance, a public comment was submitted to a federal agency requesting that the staff consider the cumulative loss of prairie habitat in the region since initial agricultural settlement in the 1800s. Construction of the proposed facility would not have impacted any remaining prairie vegetation, although it would have resulted in the loss of several hundred acres of forest and cropland that formerly supported prairie, and could potentially be restored to prairie in the future. Forcing the agency and its license applicant to do a rigorous CIA on prairie losses dating back to

1800 is clearly beyond the intended scope of the NEPA, although briefly acknowledging the historical abundance of prairie and its near total absence at the present time would help to describe the context of a land-scape whose original vegetation has been highly altered since original settlement. Minor losses of a few native plant specimens, even if not for-mally protected under federal or state regulations, might be interpreted as more significant in such a context than if the time baseline only went back as far as a few years or even decades. The applicant withdrew its applica-tion before the agency could begin its analysis.

Time domain

The effect of a specific project may end abruptly or diminish slowly with time. The time frame for a project-specific analysis normally does not extend beyond the point where the project-specific effects diminish below the threshold level of significance. However, this same practice may not necessarily extend to the problem of assessing cumulative impacts.

Recall that the NEPA regulations define a cumulative impact to be the "...incremental effect of the action when added to other past, present, and reasonably foreseeable future actions" (§1508.7). Defining the appropriate time frame over which a CIA should be performed is often more difficult than establishing the corresponding spatial domain. The time frame of a project-specific analysis may be helpful in determining how far into the future to project the CIA.

For example, if the project impacts would extend 7 years into the future, this same time frame might in some instances also be suf-ficient for performing the CIA. It is not uncommon, however, to find that the time frame must be expanded beyond that for the project itself. Figure 1.1 shows a project's direct and indirect impacts diminishing until a point is reached, approximately 13 years into the future, at which these effects drop below the point of significance.[10] The analysis of direct and indirect impacts would normally not extend beyond this point in time. As shown in Figure 1.1, one or more future actions affecting this environmental resource would be triggered around the 16th year into the project.[11] The effects of this future action(s) increase over time. The cumulative impact is, therefore, the summation of the dissipating project-specific impact and the increasing effect of this future action(s). As shown in this figure, the cumulative impact slowly increases until it finally breaches the significance threshold approximately 30 years into the future. Consequently, the time frame over which the cumula-tive impact must be evaluated is substantially greater than that for the project-specific impacts alone.

A potential time constraint on the CIA time frame simply involves not extending the analysis beyond the point in which the impacts of the reasonably foreseeable future actions can be identified or meaningfully

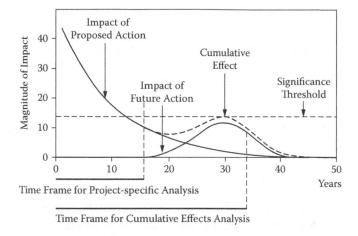

Figure 1.1 Example of a time frame for a cumulative impact assessment. (Courtesy Council on Environmental Quality.)

evaluated; however, such a cutoff point must be defensible, that is, the analysts should be prepared to demonstrate that these future impacts could not reasonably be identified or defined, or meaningfully evaluated.

Spatial domain

With respect to direct impacts, it is usually sufficient to limit the spatial bounds of the analysis to the immediate area in which the project would occur. The spatial domain used in the analysis of indirect effects frequently needs to be expanded beyond the bounds used for the analysis of direct effects. For a CIA, the geographic bounds may have to be expanded beyond that which is deemed sufficient for evaluating either the direct or indirect effects.

Choosing the appropriate spatial domain for a CIA is critical and depends on the nature of the proposal and the potentially affected environmental resources. A cumulative boundary may involve considering an entire human community, groundwater system, airshed, watershed, ecosystem, or a basin.

Available guidance on how to define the geographic bounds of an impact analysis, whether for direct and indirect or for cumulative impacts, is limited and largely left to the professional judgment of the analysts. The geographic area of interest is a factor relating to context. Recall that the CEQ NEPA implementing regulations define "context" as (40 CFR §1508.27)

> *Context. This means that the significance of an action must be analyzed in several contexts such as society as a whole (human, national), the affected region, the*

affected interests, and the locality. Significance varies with the setting of the proposed action. For instance, in the case of a site-specific action, significance would usually depend upon the effects in the locale rather than in the world as a whole. Both short- and long-term effects are relevant.

The EPA in 1999 guidance on cumulative impact assessment states the following:

EPA reviewers should determine whether the NEPA analysis has used geographic and time boundaries large enough to include all potentially significant effects on the resources of concern. The NEPA document should delineate appropriate geographic areas including natural ecological boundaries, whenever possible.

The reason the EPA does this is to ensure that analysts consider a broad enough plate of related actions whose impacts could additively or synergistically combine with the proposed action to result in cumulative impacts greater than those resulting from the proposed action alone. The EPA goes on to say in the same document:

Spatial and temporal boundaries should not be overly restricted in cumulative impact analysis. Agencies tend to limit the scope of their analyses to those areas over which they have direct authority or to the boundary of the relevant management area or project area. This is often inadequate because it may not cover the extent of the effects to the area or resources of concern.

Establishing threshold levels

The criteria for judging the significance of cumulative effects are not different from those for direct and indirect impacts, but thresholds and irreversible changes in the use of critical resources will frequently be of greater concern. A *threshold* is defined in *Webster's Dictionary* as "the point at which a physiological or psychological effect begins to be produced." Emphasis on identifying thresholds often has its roots in the desire to allow some project(s) to proceed until the magnitude of the total effect reaches a critical point at which some type of restriction or regulation is necessary to control the impacts of future actions.

There are no commonly accepted definitions or criteria as to what constitutes an acceptable threshold level. Such threshold levels are often based on some perceived level of impact or risk. Moreover, such perceptions are often based on limited data concerning the spatial and temporal bounds (time frame) of the impact.

Such thresholds are often not physically or biologically based. Instead, a point is reached at which the public or regulators become sufficiently alarmed to demand environmental protection controls. The level of an acceptable impact often has two components: environmental and social.

Thresholds may be expressed in terms of targets, goals, guidelines, or standards, carrying capacity, or limits of acceptable change, each term reflecting different combinations of scientific data and societal values. A threshold, for example, might involve a maximum concentration of a pollutant beyond which health may be adversely affected, or perhaps a maximum number of square miles of natural undisturbed land cleared before visual impacts become unacceptable to the public. A temptation exists to use regulatory thresholds triggering permit requirements, such as maximum concentration levels (MCLs) established under the US Clean Water Act. However, regulatory thresholds are not always grounded in physical or biological fact but rather in considerations necessary to reach political compromise.

Ideally, if the combined effects of all actions within a region do not exceed a certain limit or threshold, the cumulative effects of an action might be interpreted as acceptable (i.e., nonsignificant). In practice, however, cumulative effect assessments are often hindered by a lack of such thresholds. In the absence of defined thresholds, one might (1) suggest an appropriate threshold; (2) consult various stakeholders, government agencies, and technical experts; or (3) acknowledge that there is no threshold, determine the cumulative effect and its significance, and let the decision maker decide whether a threshold is being exceeded.

Determining the scoping of actions to evaluate

To ensure the inclusion of resources that are most susceptible to degradation, cumulative impacts can be anticipated by considering where cumulative effects are likely to occur and what actions would most likely produce cumulative effects. In initiating a cumulative impact analysis, practitioners should[12]

- Determine the area that would be affected
- Make a list of the resources within that zone that could be affected by the proposed action
- Determine the geographic areas occupied by those resources outside the project impact zone

Identifying present and future activities to include in the CIA

Past, present, and reasonably foreseeable future activities for inclusion in a CIA can be readily identified once the temporal and spatial boundaries of a project's effects are identified. One rule of thumb is that if the effects of known or reasonably foreseeable future activities overlap with those of the proposal in either space or time, they should normally be included in the CIA. Analysts should consult with nearby industrial operators and incorporate information about their anticipated growth into the CIA. A reasonable attempt to gather information must be demonstrated.

One school incorrectly holds that uncertainty can be avoided by including only those projects and activities known with certainty. However, such a simplistic approach almost certainly underestimates cumulative effects by neglecting the current understanding of what is reasonably foreseeable. Environmental forecasts of this kind are of limited value because they anticipate the lower bounds of plausible future conditions.

Reasonably foreseeable activities can be viewed as those that are ongoing and are likely to continue, and those that can be anticipated as a result of external trends, such as increasing tourism. Municipal planning and zoning offices are usually the best source of information on projects that are contemplated or under review for a given region. Reasonably foreseeable projects should generally include the categories of proposed activities denoted in Table 1.2 unless there is a particular reason to exclude them.

The last category of reasonably foreseeable future activities requires some further explanation. If another project could not take place without the reviewed project also taking place, it may be considered a "connected action." Connected actions must be evaluated together in the same NEPA document. Impacts from both actions should be evaluated as direct and indirect impacts resulting from the overall sequence of actions. The sequence of actions is a single combined proposed action with a single set of direct or indirect impacts. The CEQ considers actions connected if one

Table 1.2 Categories of Reasonably Foreseeable Future Activities that Should be Included in the Cumulative Impact Assessment

---

- Projects officially announced by a proponent
- Projects identified in a development plan (such as a comprehensive plan or master plan) for the area
- Projects not directly associated, but which would likely be induced as a result of the project's approval
- Projects that have been formally approved
- Projects currently undergoing regulatory review with a reasonable possibility of approval
- Projects directly associated with the project under review

---

is automatically triggered by the other, if one cannot proceed unless the other is taken previously, or if the individual actions are part of a larger action and depend on the larger action for their justification. Only if the associated action has "independent utility" can its impacts be separated out as cumulative impacts.

A specific rationale should be given for excluding future projects and activities that appear to fit one of the above categories. Beyond these reasonably foreseeable categories of activities are more speculative possibilities; such speculative projects may or may not be appropriate for inclusion in the CIA.

Considering related actions

In one case, an agency was sued for preparing individual EAs on separate mining claims that involved a cumulatively significant impact. The court concluded that an EIS was necessary when a number of related actions cumulatively have a significant environmental impact, even if the separate actions alone would not. In the words of the court, "...once the cumulative impact of a number of mining claims crosses the threshold of [a] significant effect on the environment, a discussion of those cumulative effects in individual EAs no longer complies with NEPA."[13]

Actions on private lands Case law indicates that unrelated actions on private lands must still be considered. In one case, the Eighth Circuit court ruled that an EA must consider the impacts of activities reasonably expected to occur on private lands.[14]

Considering connected actions

In 1988, the US Forest Service was challenged for preparing nine separate EAs on connected actions. In reviewing the case, the court found that the plaintiffs had raised serious questions as to whether these timber sales would result in a cumulatively significant impact. The court found that the agency's findings of no significant impact (FONSI) were inappropriate because the EAs did not adequately address connected actions and the cumulative effects of proposed and contemplated actions. The court concluded that the scope of these connected actions was broad enough so as to require preparation of an EIS.[15]

In the same year, another proposed action was challenged on similar grounds. The EA failed to evaluate the cumulative effects of connected actions involving reconstruction of a 17-mile segment of a 70-mile road, as well as other segments of the road reconstruction project, related timber sales that justified the entire project, and other reasonably foreseeable future actions. The court found that the connected actions, in addition to other reasonably foreseeable future actions, could result in a cumulatively significant impact. This was because there was an inextricable nexus between the logging operations and the road construction. The court concluded that the EA failed to evaluate the ongoing and future timber harvest and the road reconstruction.[16]

One of the most important cumulative impact cases involved the US Fish and Wildlife Service. The agency had prepared other independent documents indicating that related and cumulative impacts might be leading to aquatic habitat degradation. Such degradation was unaccounted for in the individual EAs that it had prepared. The court found that lack of an overall effort to evaluate cumulative impacts could result in detrimental effects on the recovery of the wolf population. This was sufficient to raise serious questions regarding whether the road and the timber sales would result in a significant cumulative impact. The agency was ordered to prepare an EIS to analyze such effects.[17]

Induced growth

Induced actions are activities that will reasonably follow or be triggered by approval of the proposal. These actions may not be officially announced or be part of any official plan. Induced actions often are non-federal actions not subject to the NEPA review process. They simply happen, and practitioners must examine their likelihood, based on existing use, precedent, and implementation of the proposal. Often they have no direct relationship with the action under assessment and simply represent the growth-inducing potential of an action. For example, a new road might spawn later service roads, increased recreational activities, hunting, fishing, and construction of new service facilities such as gas stations. A practitioner can usually only conjecture as to what they may be, their extent, and their environmental implications. When combined with highly successful mitigation measures, proponents may confidently claim that there are no cumulative effects. However, induced actions may represent the only source of important future actions contributing to cumulative effects.

Disregarding future actions

Future actions can generally be disregarded if[18]

- They lie outside the geographic boundaries or time frame established for the cumulative effects analysis.
- They will not affect resources that are the subject of the cumulative effects analysis.
- Their inclusion in the analysis is considered to be arbitrary (i.e., lacks a logical basis for inclusion).

The courts have struggled with the problem of determining when future actions can be disregarded as "remote or speculative" versus those that must be analyzed. In one case, a court concluded that an EA prepared for mining operations did not need to consider the cumulative impact of other planned mines. This decision was premised on the fact that there was no practical commitment to future mining operations. The court concluded that a NEPA analysis must generally consider impacts of other proposals "…only if the projects are so interdependent that it would be unwise or irrational to complete one without the others."[19]

Identifying impacts of past actions

From a theoretical perspective, the consequences of an act may go forward into eternity, and the causes of an event can go back to the beginning of human events. Some reasonable boundary must be established for determining the limits of such an analysis. But how does one establish a defensible boundary for determining the beginning and end of the cumulative impact analysis?

Early court direction for assessing impacts of past actions

Determining what constitutes an adequate assessment of the impacts of past actions has been particularly tortious, providing fodder for numerous lawsuits. Described below is a brief history of how this issue has evolved. This section also provides practitioners with a practical and defensible method for simplifying their CIAs.

Early court direction The historic case of *Fritofson v. Alexander* focused on the specific criteria that must be addressed to perform an adequate CIA.[20] With respect to performing an adequate cumulative effects analysis, the court laid out five requirements that an adequate CIA must meet. Among these requirements was the burden of identifying "other actions—past, proposed, and reasonably foreseeable—that have had or are expected to have impacts in the same area." Many viewed this direction as opening a Pandora's box of complexity in terms of preparing an analysis that met the court's criteria.

A subsequent case, *Lands Council v. Powell*, reinforced the direction provided in *Fritofson v. Alexander*. Specifically, the court sought to provide guidance on the practice of cumulative impact assessment involving a watershed restoration project related to a timber harvest.[21] Among other things, the court defined an adequate CIA as

1. Separately discussing "prior harvests from different projects … as to the consequences of each;" and
2. To evaluate cumulative effects of past timber harvests, the EIS needed to provide adequate "data of the time, type, place, and scale of past timber harvests and should have explained in sufficient detail how different project plans and harvest methods affected the environment."

Essentially, this case directed the agency to individually analyze the incremental effects of all past harvests, which amounted to hundreds of actions in a very large watershed over the past 60 years, and summate them to arrive at the current conditions that would be affected by the proposed action and other reasonably foreseeable future actions.

CEQ's guidance on assessing impacts of past actions

The implications that such direction would have on other CIAs could be impractical and overwhelming. In practice, such direction could have the effect of turning many CIAs into a virtually hopeless exercise. In response to the *Lands Council* decision, the CEQ published a memorandum to federal agencies entitled "Guidance on the Consideration of Past Actions in Cumulative Effects Analysis."[22] The CEQ guidance provides that

> ...*generally*, agencies can conduct an adequate cumulative effects analysis by focusing on the current aggregate effects of past actions without delving into the historical details of individual past actions.

The memorandum added that

> ...agencies are not required to list or analyze the effects of individual past actions **unless** such information is necessary to describe the cumulative effect of all past actions combined.

Thus, in the CEQ's opinion, an agency may eliminate the specific listing of individual past actions, if it can otherwise describe the existing condition of a resource, even though the past actions may have caused the current condition. Essentially, the memorandum is advising practitioners to focus on the present effects of past actions, instead of the *past actions* themselves. Nevertheless, the memorandum recognizes that specific data on past actions is sometimes useful, as information about direct and indirect effects of individual past actions may aid in forecasting the direct and indirect effects of the proposal at hand.

Therefore, in the earlier example regarding historical prairie losses in the region of a proposed facility, the agency would not be expected to inventory the individual prairie loss events dating back to the 1800s—which would be completely impossible even if the exercise could possibly contribute useful information. Simply acknowledging the nearly complete absence of prairie in the landscape at the present time would be adequate.

Court direction validating CEQ's interpretation of past actions
Despite the CEQ's use of caveats such as "generally" and "unless," the US District Court in the Eastern District of Washington, deferred to the CEQ's guidance in the case of *Conservation Northwest v. USFS.* This court held that an agency did not need to catalogue and individually analyze the effects of past timber activities in a post-fire salvage sale proposal. The court concluded that[23]

> ...*the EA's cumulative analysis compliant with CEQ's recent publication, Guidance on the Consideration of Past Action[s] in Cumulative Effects Analysis ... CEQ was created by NEPA and is the body responsible for promulgating NEPA's implementing regulations... .As a result, CEQ's interpretation of NEPA is entitled to substantial deference.*

Several courts, in particular the Ninth Circuit, have likewise deferred to the CEQ's guidance that past actions need not be individually catalogued and analyzed.[24] Based on a review of case law, Magee and Nesbit generally agree with the CEQ's interpretation; they report that an agency generally prevails:[25]

> ...*where it provides its own detailed methodology for analyzing cumulative impact significance, even though different than that described in Lands Council. The NEPA documents that fail to describe their methodology or to refer to CEQ's past actions guidance and make a convincing finding that "listing or analyzing the effects of individual past actions is not necessary to describe the cumulative effect of all past actions combined" are vulnerable to challenge regarding the adequacy of their cumulative impact analyses.*

Defining a defensible baseline for assessing impacts of past actions
In 2004, the Supreme Court heard a case in which opponents charged that an EA failed to take into account the environmental effects of increased cross-border operations of Mexican motor carriers.[26] The Supreme Court did not agree with the opponent's arguments; in reaching its finding, the court cited a ruling from *Metropolitan Edison Co. v. People Against Nuclear Energy*[27] that "NEPA requires 'a reasonably close causal relationship' between the environmental effect and the alleged cause. The [prior] Court analogized this requirement to the 'familiar doctrine of proximate cause from tort law.'"[28]

Doctrine of proximate cause   In placing limits on the extent of an analysis, the Supreme Court appears to have been saying that NEPA requires "a reasonably close causal relationship" between an impact and its cause.[29, 30] This ruling has its roots in the familiar doctrine of proximate cause from tort law.[31] In other words, the Supreme Court interpreted NEPA's cumulative impact provision as requiring agencies to only consider the incremental impact *proximately* caused by the proposed action in the context of the existing conditions, together with other present and future actions having an effect on the same resource.

In placing limits on the extent of an analysis, the Supreme Court cited a ruling that NEPA requires "a reasonably close causal relationship" between an impact and its cause.[32] This ruling has its roots in the familiar doctrine of proximate cause from tort law. The concept of *proximate causation* from tort law provides a defensible concept for limiting the extent to which an agency must consider potentially affected resources and is defined as

> *"Proximate cause" is merely the limitation which the courts have placed upon the actor's responsibility for the consequences of the actor's conduct. In a philosophical sense, the consequences of an act go forward to eternity, and the causes of an event go back to the dawn of human events, and beyond. ... As a practical matter, legal responsibility must be limited to those causes which are so closely connected with the result and of such significance that the law is justified in imposing liability. Some boundary must be set to liability for the consequences of any act, upon the basis of some social idea of justice or policy.*[33]

Example of the proximate cause test   Frequently the causal chain of events involves an overly *convoluted or speculative assessment* that may not be realistic or defensible. For example, consider an action to control encroachment of non-native invasive plants in a national forest. The NEPA analysis traces the chain of events from application of herbicides to health effects miles away. Killing vegetative cover might allow the herbicides to be carried downslope by runoff. The contaminants can eventually flow into a stream. After traveling downstream, the contaminated water can infiltrate into soil. Plants can take up the contaminated water into their roots and the contaminants can lodge in the plant tissue. Grazing animals can eat the contaminated plants. The timeline for this transit may lie beyond the published toxicity half-life of the herbicide. While such a chain of events may be remotely possible, it is outside the realm of reasonable probability. Such a chain of events may not provide a "reasonably close causal relationship" and therefore may well fail the court's proximate cause test.

A five-step procedure for accessing cumulative impacts

The following process is modified after a paper published by Magee and Nesbit that proposed the following process for performing a defensible CIA.[34] The process can broadly be viewed as a three-phase procedure:

A. Determine whether a proposal would "proximately" cause a significant effect on a given resource.
B. Investigate how the affected environment has reached its current condition and projected trajectory as a result of the aggregate effects of past actions (i.e., the affected environment) and define an appropriate spatial and temporal scale for the impact assessment.
C. Determine what happens to the condition and trend of this resource if no action is taken, including effects of other present and reasonably foreseeable future actions (i.e., the no-action alternative trajectory) and what happens to this maintained or altered trend if the proposed or alternative actions are taken (i.e., the incremental and resulting cumulative effects of the action alternatives).

Using the *proximate cause test* as an early analytical step reduces unnecessary descriptions of what may later prove to be "unaffected environments." This step involves tracing the *causal chain of events* (cause and effect) from an action under investigation to an impact on a particular resource to determine if a *reasonably close relationship* exists, as opposed to some *remote or speculative causation*. In sum, this test determines whether an action would reasonably and foreseeably cause a measurable or important impact on a resource of concern and limits an agency's analysis to those resources thus affected. The *proximate cause test* is used as the principal basis for the five-step CIA procedure described shortly. Before delving into this procedure, it is instructive to consider the concept of no action.

No-action baseline

As explained in Chapter 6, the no-action alternative does not necessarily infer that "nothing happens." Even under the no-action alternative, the affected environment may change or evolve (population increase and cities expand; roads deteriorate and traffic may increase; water consumption increases around nearby residential areas; non-native invasive species encroach into new areas; vegetation and trees continue to grow). Present and future actions independent of the proposed action and action alternatives (i.e., no-action alternative) may alter a resource's condition and trend from that created by past actions. The reader is directed to the author's (Eccleston) companion text, *NEPA and Environmental Planning: Tools,*

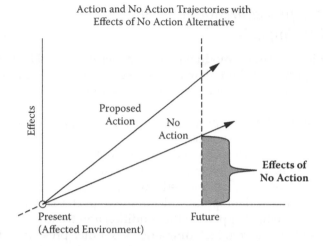

Figure 1.2 In this simplified diagram, the effects of the no-action alternative are represented as a projection or trend of a resource's condition over time (determined by the continued influence of past actions and any new effects from other present and reasonably foreseeable future actions). In this simplified diagram, the effect is depicted as trending linearly upward. (Courtesy of the National Association of Environmental Professionals. Magee, J., and Nesbit, R., *Journal of Environmental Practice*, 10(3), 107–115, 2008.)

Techniques, and Approaches for Practitioners for a more thorough treatment of taking no action.[35] The term "no-action trajectory" is used to indicate that the condition of the affected environment is likely to change even if the proposed action or one of its action alternatives is not pursued.

Figure 1.2 illustrates how present and future actions under the no-action alternative may maintain or alter a resource's condition and trend from that created by past actions.

Five-step procedure

This section describes a five-step process for performing a CIA (see Figure 1.3). Most NEPA assessments begin by describing the affected environment prior to evaluating the environmental consequences, and the discussion of environmental consequences typically leads off with the no-action alternative. A more efficient procedure, however, may be an iterative process that does not proceed in such a linear fashion. The five-step sequence of analysis described below necessitates that the environmental investigation be performed in a manner distinctively different from the way in which the analysis is typically presented in the NEPA document. Applying the proximate cause test during the first step prevents unnecessary analysis, effort, and descriptions of what might

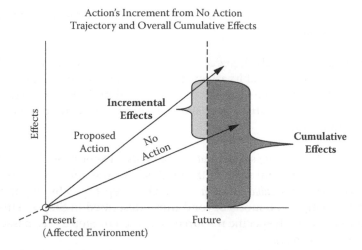

Action's Increment from No Action
Trajectory and Overall Cumulative Effects

Figure 1.3 In this simplified diagram, the incremental direct and indirect impacts are shown as the change from the no-action alternative trajectory. The cumulative impact is represented as the proposal's incremental effects in combination with the effects of other present and reasonably foreseeable future actions under the no-action alternative. (Courtesy of the National Association of Environmental Professionals. Magee, J., and Nesbit, R., *Journal of Environmental Practice*, 10(3), 107–115, 2008.)

later prove to be "unaffected environments" relative to the proposed action; it may also prevent irrelevant analyses of impacts from other actions occurring under the no-action alternative.

Thus, the CIA, first determines whether a contemplated action could *reasonably* and *foreseeably* affect a given resource. While this may appear to be a seemingly instinctive step, it avoids expending needless effort in analyzing past, present, reasonably foreseeable future actions for resources that are not affected, either directly or indirectly, by the proposal (proposed action or one of its reasonable alternatives). Step 1 can also be applied to the analysis of direct and indirect effects, in addition to the CIA.

Step 1: Proximate cause screening test All potentially affected resources are screened to determine if they could be proximately affected by the direct and indirect effects of the proposal. Analysts trace the causal chain of events (cause and effect) from the proposal to a particular resource under investigation. The proximate cause test is used to determine whether an action would reasonably and foreseeably cause a measurable or important impact on the resource. The essence of this test is determining *whether there is a reasonably close causal relationship between an action and the resulting impact.* Any resource that is not proximately affected in an important way can be dismissed from the CIA, as well as the assessment of direct and indirect impacts.

Step 2: Describe the resource   Once it has been determined that a resource may be reasonably affected (directly or indirectly) by the proposal, then that resource's current condition and projected trend (derived from *past actions and natural events*) are described as a component of the affected environment.

Step 3: Determining spatial and time bounds   Determine the appropriate spatial and temporal scale of the analysis for the effects of *present and other reasonably foreseeable future actions*. This step typically expands the spatial and temporal bounds beyond those described in Step 2. This step is performed on each specific resource identified in Step 2 (i.e., passed the proximate cause test).

The analysis can be "bounded" by applying a principle from CEQ's cumulative effects handbook: "evaluating resource impact *zones* and the life cycle of *effects* rather than projects."[36] A defensible justification should be provided as to how the spatial and temporal bounds of the CIA were determined.

Step 4: Determine no-action baseline   Determine what other recent and reasonably foreseeable future actions, not including the proposal, would affect each resource within the same spatial and time bounds designated in Step 3. The reasonably foreseeable future actions include other planned or anticipated actions (not spawned or triggered by the proposal) that are independent of the proposal. Specifically, how would the environmental baseline (no-action trajectory) change for each resource as a result of taking no action (see earlier section that explains the no-action baseline). This no-action baseline represents the impacts of past, present, and reasonably foreseeable future actions, neglecting the direct and indirect impacts of the proposal. The results of this investigation (both the actions and their effects) are documented in the section that describes the no-action alternative.

Step 5: Determine proposal's incremental impact   The final step involves combining the direct and indirect impacts of the proposal with the effects of all other past, present, and reasonably foreseeable future action (no-action baseline described in Step 4). Essentially, this step assesses how the proposal's direct and indirect impacts would incrementally change (i.e., add to, subtract from, or synergistically interact with) the no-action baseline described in Step 4. The description of the overall effect (e.g., incremental effect of the proposed action with the combined impacts of other past, present, and reasonably foreseeable future actions) constitutes the CIA.

Managing and performing a cumulative impact analysis

The purpose for preparing the NEPA analysis is to predict potential adverse impacts of a proposal and to design remedies (alternatives or mitigation) that can prevent or in some other way lessen the impact. The purpose of a CIA is essentially the same. However, the greater analytical complexity, scale, and inherent uncertainties associated with predicting future activities tend to make a CIA substantially more difficult and substantially less precise.

A principal concern involves determining the minimum degree of analysis and data requirements that will support a defensible and robust cumulative impact prediction. This concern is greater for EAs than for EISs. EAs have generally been more likely than EISs to ignore a CIA or contain an inadequate CIA. It is often beyond the scope of many EAs to include a detailed, quantitative CIA. However, most EAs should include at least a qualitative assessment of the potential for cumulative effects. As is broadly true for NEPA analyses, especially EAs, the CIA should generally be of greatest rigor for that issue or those issues with the greatest potential for significant impacts.

Components of an adequate cumulative impact analysis

A cumulative effects study must identify[37]

- The specific area in which effects of the proposed project would be felt
- Impacts that are expected in that area from the proposal
- Other past, proposed, and reasonably foreseeable future actions that have had or could be expected to impact the same area
- Expected impacts from these other actions
- Overall expected impact if the individual impacts were allowed to accumulate

In one case, a court deferred judgment in favor of an agency, finding that it had considered effects of a timber sale in the context of past and reasonably foreseeable logging; the agency had constructed mathematical models and had performed extensive field investigations to calibrate and verify its models. It had also actively sought public comment.[38]

The demanding requirements of a CIA (i.e., combining effects, expanded spatial and temporal boundaries, and perhaps resource sustainability) can best be addressed by developing a conceptual model and using a modern suite of tools such as geographic information systems (GISs), questionnaires, interviews, and panels; computer models; checklists; matrices; networks and systems diagrams; and trends analysis.

Data

It is not generally advisable to embark on costly data collection and analysis without careful consideration of the results. Analysts often adopt a "coarse filter" approach to data collection; that is, the level of information is not as detailed as in a standard analysis of the direct/indirect effects because of the much larger bounds of the CIA. For example, vegetation field studies for the assessment of direct and indirect effects may be relatively intensive within a proposed construction footprint and involve on-site plot sampling and mapping. However, for a regional CIA study involving hundreds of square miles, the analysis may have to be based on coarser satellite imagery or existing vegetation surveys completed at very broad scales. For example, an EIS for a proposed plant used on-site observational data to describe affected vegetation on the project site but used state land cover maps based on satellite imagery to characterize vegetation in the surroundings.

The setting of spatial and temporal boundaries will be at least partly a function of data availability and the level of uncertainty and confidence in making environmental predictions.

Assessing cumulative effects

Many approaches have been advanced for performing a CIA. In one fashion or another, analysts typically determine the separate effects of past actions; present actions; other reasonably foreseeable future actions; and the proposed action (and reasonable alternatives). Cumulative effects can be calculated once each group of effects has been determined. For example, with respect to air quality, one approach might involve evaluating all existing emission sources for which prevention of significant deterioration (PSD) permits are in the process of being reviewed or approved, and those for which a PSD permit is planned but has not yet been submitted. The combined emissions create an effect on air quality, the significance of which can be determined by comparing the cumulative concentration of pollutants emitted to threshold concentrations specified in the National Ambient Air Quality Standards (NAAQS).

Once the impacts resulting from past, present, and reasonably foreseeable future air quality actions have been combined together (i.e., cumulative impact baseline), the analysts may then "add" the impact of the proposed action to better understand how it affects the cumulative impact baseline. Once these effects have been determined, a table can be used to organize and present itemized effects into categories of past, present, proposed, and future actions, and the resulting cumulative effect. Table 1.3 shows how such tables can be constructed. The table compares a narrative versus a quantitative description of the cumulative effects associated with an increase in NO_x concentrations.

Table 1.3 How a Table Can Summarize a Cumulative Impact Analysis

	Past action	Present action	Proposed action	Future action	Cumulative effect
Narrative description	No discernible effect on NO_x levels	Notable deterioration in visibility during spring, but standards are met	Visibility further affected by the project, but standards are met	Increased vehicle emissions expected	Standards likely to be exceeded
Quantitative assessment		5% increase in NO_x concentration, but standards are met	10% increase in NO_x concentration, but standards are met	5% increase in NO_x concentration	20% increase in NO_x concentration will exceed regulatory standards

The following 16-step approach is recommended for assessing cumulative impacts (Table 1.4). The steps depicted in Table 1.4 are common to most CIAs.

For supplemental guidance on performing a CIA, the reader is referred to the CEQ's publication entitled *Considering Cumulative Impacts*, which describes how to define the scope of a cumulative analysis, define the spatial or temporal baseline, and assess truly cumulatively significant issues.[39]

Evaluating cause-and-effect relationships

As is true for assessing direct and indirect environmental impacts from an individual action or alternative, determining how a particular resource responds to an environmental disturbance is essential in determining the cumulative effect of multiple actions. Analysts must therefore gather information about cause-and-effect relationships. Once all the important cause-and-effect pathways are identified, the analyst determines how environmental resources respond to a potential disturbance. Cause-and-effect relationships for each resource are used in computing the cumulative effect resulting from all actions considered.

Typically, analysts will determine the separate effects of the proposed action, and other past, present, and reasonably foreseeable future actions. The cumulative effect can then be summated once each group of effects is determined.

Performing a qualitative analysis   A cumulative impact analysis is sometimes limited to a qualitative evaluation because the cause-and-effect relationships are poorly understood or difficult to accurately quantify. In still other cases, there may not be sufficient site-specific data available to permit

Table 1.4 A 16-Step Approach for Performing a Cumulative
Impact Assessment

Key impact assessment components	Cumulative impact analysis steps
Scoping	1. Identify the significant cumulative impact issues associated with the proposed action and define the assessment goals.
	2. Establish the defensible geographic and time frame scope for the CIA. Using the Proximate Cause Test, determine defensible direct or indirect effect bounds of the analysis (i.e., a reasonably close causal relationship rather than a remote or overly speculative chain of causation).
	3. Identify other past, present, and reasonably foreseeable future actions affecting the resources, ecosystems, and human communities of concern that must be considered in the CIA.
Describing the affected environment	4. Characterize the resources, ecosystems, and human communities identified during scoping in terms of their response to change and capacity to withstand stresses.
	5. Characterize the stresses affecting these resources, ecosystems, and human communities and their relationship to regulatory thresholds or other applicable threshold values.
	6. Define and describe the baseline conditions (affected environment) for the resources, ecosystems, and human communities.
Determining the environmental consequences	7. Identify and define environmental disturbances (e.g., emissions, effluents, noise, waste) produced by past, present, and reasonably foreseeable future actions that could affect the resources, ecosystems, and human communities.
	8. Identify the important cause-and-effect relationships in which these environmental disturbances would affect human activities and resources, ecosystems, and human communities (e.g., how would the environmental disturbances affect humans and environmental resources?).
	9. Determine a defensible spatial and temporal bound of analysis for the effects of other present and reasonably foreseeable future actions, based on the temporal and spatial *effects* of the proposal on each resource of concern identified in Step 2.
	10. Combine or "add" the effects of the past, present, and reasonably foreseeable future actions together to produce a "cumulative impact baseline" for the resources, ecosystems, and human communities that could be significantly affected.

Table 1.4 A 16-Step Approach for Performing a Cumulative
Impact Assessment (Continued)

Key impact assessment components	Cumulative impact analysis steps
	11. Combine or add the impacts of the proposal to the cumulative impact baseline to determine how the project would affect the cumulative impact baseline.
	12. Determine the magnitude and significance of the cumulative effect.
	13. Modify or add alternatives or mitigation measures to avoid or reduce the cumulative effects.
	14. Include an uncertainty analysis that discusses areas of uncertainty and potential errors.
	15. Monitor the cumulative effects of the selected alternative.
	16. Consider using an environmental management system or adaptive management approach for monitoring and addressing any impacts that exceed original projections.

a quantitative analysis. Faced with such constraints, analysts may want to consider performing a qualitative analysis in a manner similar to that shown in the following example (Table 1.5). If no numbers are available, the analyst may categorize the magnitude of effects using qualitative descriptors such as "extensive," "broad," or "localized." Contrast this example with a cumulative impact assessment using quantitative descriptors depicted in Table 1.6.

Dealing with uncertainty

Arguably, the largest source of uncertainty is the result of imperfect knowledge of baseline conditions and present activities, limited understanding of the primary and indirect impacts of activities, more complicated interactions, and uncertainties about future development scenarios. A good CIA should include an uncertainty analysis that qualitatively identifies and discusses each individual source of uncertainty encountered throughout the analytical process as well as the overall "summed" uncertainty. The evaluation of overall uncertainty needs to acknowledge how uncertainty propagates throughout a sequential analytical process such as CIA. Uncertainty associated with later steps in the process is magnified by the underlying uncertainty inherent in the earlier steps that form the foundation of analysis.

One means of addressing this uncertainty is through monitoring. Any monitoring program should include a

- Description of how project implementation can be monitored
- Plan to respond to unfavorable outcomes

Table 1.5 Cumulative Effects Analysis Using Qualitative Descriptions

Environmental resources	Past actions	Present actions	Proposed actions	Future actions	Cumulative impacts
Vegetation	Widespread; most presettlement forest converted to farmland or successional forest; most climax forest limited to public lands (10–20% of landscape)	Localized cutting of woodlots continues on private lands	Less than 10 acres of climax forest lost	Localized cutting expected to continue in private woodlots: federal and state management plans call for keeping limited areas of climax forest on public land	Climax forest will continue to persist at close to the present level due primarily to management of public lands
Visual resources	Presettlement forest landscape converted to agricultural landscape with scattered forest patches	Little to no development activities other than occasional building of scattered individual residences	New highway is a visual intrusion into bucolic landscape	New residential and commercial development may be induced around interchanges on the new highway	Landscape will be less bucolic and more suburban in character
Wetlands	Statewide, it is estimated that roughly 40% of presettlement wetlands remain: no data for local vicinity	One individual Section 404 permits pending for 12 acres of wetland fill; minor losses of up to 0.3 acres each expected due to nationwide permitting	4.7 acres of forested wetlands and 0.2 acres of emergent wetlands will be permanently lost	No individual Section 404 permits pending; minor losses of up to 0.3 acres each year expected due to nationwide permitting	85% of presettlement wetlands lost over next 15 years

Table 1.6 Example of a Cumulative Effects Analysis Using
Quantitative Descriptions

Environmental resources	Past actions	Present actions	Proposed actions	Future actions	Cumulative impacts
Vegetation	30% of presettlement vegetation lost	1% of vegetation lost this year	3% of existing vegetation would be lost	1% of vegetation lost yearly for next 15 years	49% of presettlement vegetation lost over next 15 years
Wetlands	30% of presettlement vegetation lost	1% of wetlands lost this year	9% of existing wetlands would be lost	3% of wetlands lost yearly for next 15 years	85% of presettlement wetlands lost over next 15 years
Turtles	20% of presettlement turtles lost	2% of turtles lost this year	6% of existing turtles would be lost	3% of turtles lost annually for next 15 years	73% of presettlement turtle population lost over next 15 years

Resolving Eccleston's cumulative impact paradox

The importance of assessing cumulative impacts is underscored by one of the factors that requires consideration in reaching a determination regarding potential significance:[40]

> ...*whether the action is related to other actions with individually insignificant but cumulatively significant impacts. Significance exists if it is reasonable to anticipate a* **cumulatively significant impact** *on the environment. (emphasis added).*

A paradox, referred to as *Eccleston's cumulative impact paradox* (*Eccleston paradox*), can arise from the fact that the definition of cumulative impacts requires consideration of other impacts from past, present, and reasonably foreseeable future activities (to provide the cumulative impact baseline).[41] For a more updated and improved discussion of the following approach, the reader is referred to a recent paper by the author titled Assessing Cumulative Significance of Greenhouse Gas Emissions (*Journal of Environmental Practice*, Vol. 12, No. 2, June 2010, pp. 105–115).

The cumulative impact paradox

Approximately 30,000 to 50,000 EAs are prepared each year.[42] Many are prepared for locations or resources that have already sustained significant cumulative impacts. Yet, a Finding of No Significant Impact (FONSI), by definition, means an action that[43]

> ...will **not** have a **significant** effect on the human environment (emphasis added).

Moreover, a "categorical exclusion" (CATX) means[44]

> ...a category of actions which do not individually or **cumulatively** have a **significant effect** on the human environment and which have been found to have no such effect...and for which, therefore, neither an environmental assessment nor an environmental impact statement is required... (emphasis added).

This paradox arises from the fact that the definition of cumulative impacts requires consideration of effects from other past, present, and reasonably foreseeable future activities (cumulative impact baseline). This paradox is evidenced by the following example. Consider a proposal to construct a relatively modest federal building in a crowded downtown business area of a large city. The area has already sustained a significant cumulative impact. The downtown area is already paved over with concrete, buildings, and skyscrapers. Streets are already congested with traffic. The natural vegetation and wildlife habitat originally present in the area have been severely compromised. The visual quality of the once-bucolic setting has been significantly compromised. An adjacent stream has already been contaminated, and the water table has sustained a significant drawdown. Ambient air quality has been significantly degraded. Fish and other aquatic species in nearby rivers and streams have experienced a significant decline. Destruction of wetlands and construction of impermeable pavement have increased the risk of flooding within the city and downstream of the city. As a result of the impacts of past and present actions, a number of environmental resources have already been significantly affected, from a cumulative standpoint. Reasonably foreseeable future actions will only worsen these problems.

Consider another example involving a popular recreational campground located in a remote area. The campground serves approximately 7,000 campers per season. The campers have already extracted a heavy toll on the surrounding environment. The level of noise has already increased to the point where it is affecting some species (and many campers are, likewise, upset). The population of an exotic plant species and the habitat of a threatened species have already begun to decline. Previous campers have already affected visual resources by littering and trampling ground plants to the point where many campers are beginning to publicly complain. Finally, the water quality in a nearby stream has already been significantly degraded by unauthorized dumping of washwater by campers. The responsible federal agency would like to prepare an EA for revamping the access trail leading into the campground. Although the refurbished trail might increase the total number of campers, professional demographers have estimated (based on historical trends) that any such increase would be less than 10 to 20 recreationists per season: These additional recreationists would contribute a very small, incremental impact when compared to the 7,000 visitors already visiting the campground. Yet, as witnessed earlier, a strict interpretation of the regulations appears to preclude issuance of a FONSI under these circumstances, theoretically resulting in a full-blown EIS. The cost of preparing an EIS could actually pay for a substantial mitigation effort to repair much of the environmental damage that has already taken place. Is it reasonable to expect an agency to prepare an EIS for such a small increase in campers simply because this resource has already sustained a significant cumulative impact, even though the direct and indirect impacts of the new proposal are virtually innocuous? And would there even be tangible alternatives to refurbishing the trail?

As depicted by the aforementioned regulatory citations, a strict interpretation of the regulations implies that a CATX or EA/FONSI cannot be applied to *any* proposal that adds *any* contribution to a cumulative impact that has already breached the threshold of significance. Now, if an environmental resource has already sustained a cumulatively significant impact, how can a decision-maker declare that a proposed action contributing *any* incremental impact (however small) is eligible for a CATX or EA/FONSI? Yet, CATXs and EAs/FONSIs are routinely (i.e., incorrectly) applied across virtually all federal agencies for proposed actions, such as the two examples noted above that involve environmental resources that have already sustained cumulatively significant impacts.

Considering the NEPA's regulatory definitions and requirements, many (if not most) activities for which EAs (and CATXs) are currently prepared should actually be ineligible for a FONSI, therefore requiring preparation of an EIS; strict compliance with these regulatory provisions would result in an unreasonable and voluminous increase in the number of required EISs (even where the incremental impacts would be relatively innocuous) and, in many areas, might render the concept of a CATX/FONSI next to useless. For instance, in the example of the downtown area described above, a strict interpretation of cumulative significance could lead to the conclusion that a federal agency would have to prepare an EIS to construct something as mundane as a stoplight, walkway, or small parking lot. Clearly, a strict interpretation of the NEPA's regulatory requirements can lead to absurd, unreasonable, and politically unacceptable results. Consequently, *Eccleston's paradox* refers to

> *The illogical, unreasonable, or absurd situation in which CATXs or EA/FONSIs are routinely applied to relatively mundane actions in areas that have sustained cumulatively significant impacts, yet the application of these streamlining provisions violates the cumulative impact regulatory constraints placed on their very use.*

Importance of resolving this paradox

This paradox is much more than one of mere passing or academic interest. From a practical standpoint, how can decision makers be expected to make reasonable and consistent determinations regarding the significance of a cumulative impact when the very definition can, and frequently does, lead to absurd or unreasonable outcomes? As CATXs and EAs are far more common than EISs in the application of the NEPA, the paradox must be resolved if the analysis of cumulative impacts is to be seriously and consistently implemented in a manner that truly safeguards environmental quality.

In considering this paradox, it is important to recognize that NEPA is governed by the rule of reason. That is, "reason" should prevail when a regulatory requirement results in an absurd outcome. A regulatory provision leading to the conclusion that an EIS is required, even in situations where it would contribute little or no substantive value to the decision-making process, contradicts the rule of reason.

Moreover, the paradox also conflicts with the CEQ's direction to reduce unnecessary NEPA paperwork and delay because a strict interpretation would require that EISs be prepared for a multitude of situations where the direct and indirect impacts, as well as the incremental cumulative contribution, of an action are relatively trivial and a CATX or EA/FONSI should suffice.[45] Perhaps most importantly, it is simply *unrealistic and unreasonable to expect federal agencies to prepare EISs for many trivial or even innocuous actions merely because the existing environment or cumulative environmental baseline may already be significantly affected* from a cumulative standpoint.

Interpreting significance

As depicted in Chapter 6, the significance of an action should be considered in terms of both the *intensity* and *context* in which the impact(s) occur. Under the *significant departure principle* presented in the next section as a solution to the paradox, both the *intensity* and *context* must be considered in assessing the threshold of significance. With respect to context, the CEQ regulations state that[46]

> ...*the significance of an action must be analyzed in several contexts such as society as a whole (human, national), the affected region, the affected interests, and the locality. Significance varies with the setting of the proposed action. For instance, in the case of a site-specific action, significance would usually depend upon the effects in the locale rather than in the world as a whole.*

The regulations identify ten intensity factors that need to be considered in making a determination regarding the significance or nonsignificance of an impact.[47] As an example, one of these factors states that an agency should consider "the degree to which the proposed action affects public health or safety."[48] While ten significance factors (in addition to context) have been developed to assist agencies in assessing significance, specific regulatory direction does not exist regarding how they are to be interpreted or applied in reaching such determinations. Federal agencies have, in fact, been granted wide discretion in interpreting how such factors can be applied in reaching a determination of significance.

For example, six of the significance intensity factors states that decision makers should consider the *degree to which* a given factor may affect some environmental attribute.[49] Agencies are given no direction, however, for interpreting or determining when an impact has affected an environmental resource to such a degree that it constitutes a significant environmental impact. Such judgment is left to the discretion of the decision maker. Thus, responsibility ultimately lies with the individual decision maker for determining how such significance factors are to be interpreted. Decision makers must, therefore, exercise a considerable degree of professional judgment in making such determinations.

Significant departure principle

The following solution to this paradox is based on the fact that decision makers have been granted a wide degree of discretion with respect to interpreting significance. The *significant departure principle (SDP)* provides a cornerstone for developing an interpretation of significance that, in turn, provides a practical solution for resolving the paradox.

Would the action significantly change the cumulative impact baseline?

Under the SDP, significance, with respect to the assessment of a CIA, can be viewed in terms of the degree to which a proposed action would change (i.e., depart from) the cumulative impact baseline. In other words, significance can be viewed as the degree to which a proposed action would affect or cause a cumulative impact to change or "significantly" depart from conditions that would otherwise exist if the proposed action was not pursued.

Using the SDP interpretation, an action could be considered nonsignificant (from the standpoint of a cumulative impact) as long as it does not, together with other concurrent and reasonably foreseeable actions, cause a cumulative impact(s) to *change* or *depart significantly* from conditions that would exist if the action was not pursued. *In other words, the incremental impact is of such nonsignificance that it would not appreciably change or contribute in an important manner to the cumulative impact if it were "added" to the effects of other past, present, and reasonably foreseeable actions.* Conversely, if a proposed action, together with other concurrent and reasonably foreseeable actions, results in a substantial cumulative change to the same environmental resource (cumulative baseline), it would be deemed significant.

The paradox is therefore resolved if the significance of a cumulative impact is considered in terms of the relative change to the cumulative baseline conditions, as opposed to a strict interpretation of assessing the total combined effects from past, present, and future actions from an absolute perspective. Significance is therefore interpreted from a relative perspective, as opposed to a strict, absolute assessment. As used here, the term "absolute" denotes a strict or more standard interpretation in which the significance of an impact is assessed simply in terms of whether the threshold of significance would be or has been breached. In contrast, the term "relative" is used to denote the SDP interpretation, in which significance is assessed in terms of the relative degree of *increase* or *change* (i.e., departure) in a cumulative impact.

Consider an example where an agency needs to take an action that might result in the death of a certain species of fish. Assume that the fish and their habitat have already sustained a cumulatively significant decrease in their numbers. Assume that the action would result in the loss of ten additional fish out of the entire river. From a cumulative standpoint, this incremental loss may be considered nonsignificant if the fish species has a population numbering in the range of 100,000 (i.e., the action would decrease the fish population by 0.01%). This is true even if the population had historically been 1,000,000, prior to a 90% population reduction caused by previous agricultural and urban development of the watershed. Such a view is both justified and consistent with the rule of reason because little or nothing may be gained by preparing an EIS for an action that would not *substantially alter or change* the cumulative impact baseline, even if the environmental resource has already been significantly affected. In contrast, the same action would very likely be deemed cumulatively significant (a significant departure) if the baseline fish population in the river was only 25 in number (40% decrease in the existing fish community).

One method for implementing the SDP simply involves adding the impacts of other past, present, and reasonably foreseeable future federal and non-federal actions together to produce a cumulative impact baseline.[50] The incremental impact of the proposal can then be "added" to this baseline. The SDP is then used to interpret whether the incremental change in the cumulative impact baseline is of such importance to this environmental resource as to be deemed significant.

In applying the SDP, a review should be performed to ensure that the incremental impact would not trigger or breach a significance threshold (i.e., the straw that breaks the camel's back). If the incremental impact breaches or crosses over some critical threshold, clearly it should be considered significant. Such a threshold might involve the violation of an applicable regulatory limit or some other suitable threshold, such as a scientific or environmental constraint that, if exceeded, would destabilize a fish population. For supplemental guidance on performing a CIA, the reader is referred to the CEQ's publication entitled *Considering Cumulative Effects under the National Environmental Policy Act*, which describes how to define the scope of a cumulative analysis, define the spatial or temporal baseline, and assess truly cumulatively significant issues.[51] The CEQ has also issued a memorandum entitled "Guidance on the Consideration of Past Actions in Cumulative Effects Analysis."[52] With respect to using the SDP, this memorandum provides important guidance that may be helpful in establishing the cumulative impact baseline.

While application of the SDP would enable applicants to overlook historical impacts when deciding whether a CATX or EA/FONSI could suffice in lieu of an EIS, it would not allow assessors to ignore impacts from past actions when performing a CIA. Using the fish example noted above, the CIA would still have to consider the combined impacts of the proposed action and other present and reasonably foreseeable future action in the context of the fact that past actions had already reduced the fish population by 90%. The assessment would still have to acknowledge that minor reductions in species whose populations have already been severely depleted could have greater impact than similar reductions in abundant species. Users of the SDP would still have to consider the effects of past actions as molding the context used in a significance determination for present and future actions. The fact that a fish species had experienced a 90% population decline in the 10 years prior to the present baseline level could suggest that the species is exceptionally vulnerable to further losses of individuals, even if the cumulative loss of individuals represents only a small percentage of the remaining population. On the other hand, if the 90% loss had occurred more than a century previously, the current population might be interpreted as relatively stable, and subsequent loss of the same small percentage of the remaining population might not be significant.

Criticisms of the SDP method

A rebuttal could be raised that the SDP could allow a succession of many small projects to be implemented (without preparation of individual EISs) such that over a period of time, the cumulative incremental value of these small projects could amount to a large impact. It is conceivable that such situations might arise. As part of the CIA, however, the impacts of reasonably foreseeable future actions are expected to have already been addressed in the CIAs for the earlier projects and hence included into the cumulative impact baseline. Notwithstanding, it is simply unrealistic to expect decision makers to reach significance determinations (requiring the preparation of lengthy and expensive EISs) for mundane projects with relatively innocuous, incremental impacts, and to which reasonable alternatives may not even exist.

Decision makers and practitioners understand they must employ practical (not theoretically perfect) methods in assessing impacts. The SDP provides a more rational, objective, and practical approach for assessing significant cumulative impacts and resolving the cumulative impact paradox. Thus, the SDP offers decision makers a method for focusing on truly important cumulative incremental impacts while de-focusing attention on those small enough to be deemed unimportant to the decision-making process.

Applicability

The SDP can provide a particularly valuable tool for interpreting significance in an EA/EIS where an area or environmental resource has

1. Already sustained a cumulatively significant environmental impact, especially if during the centuries and decades prior to NEPA and modern environmental planning, but the impacts of the proposed action are so small as to contribute little or no appreciable change to the cumulative baseline.
2. Not been significantly affected and the incremental impact would not breach the threshold of significance, yet the impact might still be considered significant simply as a result of the sizeable increase in contribution to the cumulative impact baseline (this circumstance is described shortly).

Examples of the paradox

Application of the SDP is described in the following four cases involving cumulative noise quality impacts. These four examples assume that the significance threshold involves a scientifically established noise limitation.

Cases 1 and 2 represent situations where the threshold of significance has already been breached. Case 3 illustrates a situation where the cumulative impact baseline lies significantly below the threshold of significance, and the proposed action would substantially increase the cumulative baseline but would not actually breach the significance threshold. Case 4 depicts a situation where the cumulative baseline lies just below the threshold of significance, and the small contribution of the proposed action actually breaches the significance threshold.

Figures 1.4 through 1.7, correlating with each of the four cases, are included for conceptual purposes so as to clearly illustrate the concept of the paradox and the SDP; these figures are not drawn to scale. For simplification purposes, these four examples assume that the context is factored into the assessment of the significance threshold values and the assessment of the environmental impacts. Although the figures portray noise impacts, the SDP can be applied to virtually any cumulative impact.

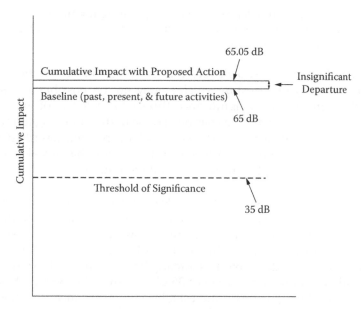

Figure 1.4 Assessing cumulative impacts, using the significant departure principle, Case 1. The cumulative significance threshold has been breached but the incremental impact is nonsignificant.

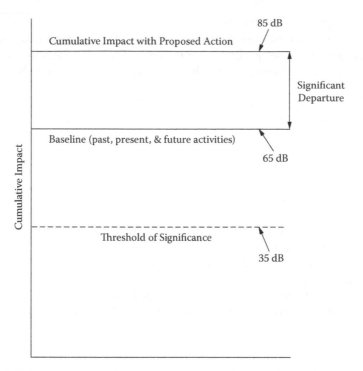

Figure 1.5 Assessing cumulative impacts using the significant departure principle, Case 2. The cumulative significance threshold has been breached and the incremental impact is significant.

The author fully acknowledges that there are many situations (perhaps the majority of cases) in which it may not be possible to assess a cumulative impact or its significance threshold in terms of an explicit number (or threshold value), as has been done in the following four cases. Nevertheless, even if an impact/threshold value cannot be described quantitatively (e.g., many visual, socioeconomic, or some biological impacts), the SDP can still be applied from a qualitative standpoint.

Case 1: Cumulative significance threshold has been breached, but incremental impact is nonsignificant   As represented by Figure 1.4, assume that the average daytime noise level (as a result of past, present, and reasonably foreseeable future actions) within a city park is 65 decibels (dB). A noise quality significance threshold has already been established for this area, which calls for a maximum level of 35 dB. Because the existing baseline level (65 dB) already exceeds the established significance threshold (35 dB), this area can be viewed as having already sustained a cumulatively significant noise impact. (The reader should note that noise intensity is measured in decibel units on a base ten logarithmic scale and thus this

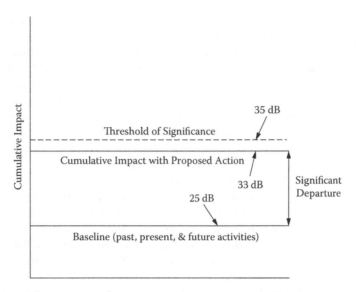

Figure 1.6 Assessing cumulative impacts using the significant departure principle, Case 3. The cumulative significance threshold has not been breached and the incremental impact is significant.

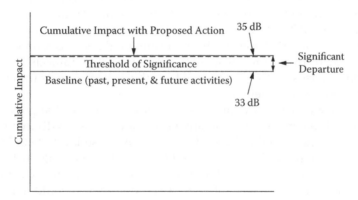

Figure 1.7 Assessing cumulative impacts using the significant departure principle, Case 4. A small incremental impact is sufficient to breach the significance threshold.

is a logarithmic rather than a linear phenomenon; human perception of loudness also conforms to a logarithmic scale.)

Assume that a proposal is under consideration to repave a potholed federal highway bordering the park. Because a federal agency has jurisdiction over the property and adjacent roads, NEPA is triggered. Assume that such a small action would typically be covered under a CATX. Further, assume that a traffic expert has concluded that as a result of this maintenance improvement project, the traffic is expected to increase slightly, raising the average daytime noise level from 65 to 65.05 dB, for a 0.05-dB increase. Because the threshold of significance has already been breached (35 dB), a *strict or "standard" interpretation* of significance suggests that any additional contribution must also be significant (regardless of how trivial an increase); an EIS could therefore be deemed necessary for this maintenance project, as well as for other, similar projects (e.g., installing a traffic light that results in a slight increase in noise as a result of stop-and-go or diverted traffic) that cause an increase in a cumulative impact baseline, regardless of how nominal the effect.

Using the SDP approach, however, the decision maker considers the proposed project from its relative incremental effect rather than from an absolute perspective. That is, he considers whether the *departure or change* in an impact(s) (rather than the *absolute magnitude* of the effect) is significant, with *respect* to the cumulative impact baseline. This involves examining the 0.05-dB incremental impact to see if it would make an important difference (i.e., change) in the cumulative impact baseline. Such an analysis involves examining the incremental impact in terms of the CEQ's ten intensity factors, as well as the context in which the incremental impact would take place.[53]

Consequently, if a 0.05-dB cumulative change is considered significant, the entire project is deemed significant and an EIS must be prepared. Conversely, if a 0.05-dB cumulative increase is considered inconsequential, posing no *significant* change to the environmental resources, an EIS is not required (with respect to consideration of cumulative impacts). The SDP, therefore, focuses on the degree to which the impact *changes* the cumulative environmental baseline.

One may consider the impacts as follows. Users of the park would be expected to consider a cumulative noise level of 65.05 dB to be objectionable, but they would also find the existing noise level of 60 dB to be equally objectionable. Park users would not be able to readily distinguish between noise levels of 60 and 60.05 dB. Therefore, the agency could justifiably conclude that the slightly increased noise level would not significantly degrade the quality of environmental conditions experienced by park users.

Case 2: Cumulative significance threshold has been breached and the incremental impact is significant   Figure 1.4 represents a case where the change is small and relatively nonsignificant (0.05-dB increase), while Figure 1.4 illustrates a case where the change is large (an increase of 20 dB, resulting in a total cumulative impact of 85 dB). Park users **would** be expected to readily notice and distinguish between a noise levels of 65 and 85 dB. The quality of environmental conditions experienced by park users would be noticeably worse following the incremental noise increase resulting from the proposed action (20 dB). The responsible agency would have to consider whether the ability of users to enjoy the park would be substantially lessened by a noise increase of 20 dB. If so, the agency would be justified in concluding that the proposal's incremental impact is cumulatively significant, requiring preparation of an EIS.

Case 3: Cumulative significance threshold has not been breached yet the incremental impact is significant Figures 1.4 and 1.5 represent are situations where the threshold of significance had already been breached prior to the proposed action. Figure 1.6 illustrates yet a third case, where the cumulative noise baseline is 25 dB, which lies well below the threshold of significance (i.e., noise level of 35 dB). In this case, however, the proposed project would increase the cumulative impact baseline from 25 to 33 dB, which is near but still below the threshold level of 35 dB. Because the proposed project would substantially increase but not actually breach the threshold of significance, a strict *interpretation* of significance might lead to the conclusion that the cumulative impact is still nonsignificant and therefore an EIS is not required.

Under the application of the SDP, however, the proposed action might more correctly be deemed to constitute a *significant departure or change* in the cumulative impact baseline, even if it would not actually breach the threshold of significance. Such a conclusion is justified because the proposed action would significantly increase the baseline (even though it has not actually exceeded the threshold of significance); a decision maker could (perhaps should) be justified in concluding that the action would constitute a cumulatively significant increase in noise impact, requiring preparation of an EIS. In this case, preparation of an EIS may be warranted because alternatives or mitigation measures may be identified that eliminate or substantially reduce the significant increase in noise level.

Consequently, the SDP can in some cases result in a determination of significance (indicating preparation of an EIS), where a more standard approach would result in the opposite conclusion (i.e., nonsignificance). The reader should note that this interpretation is consistent with one of the existing significance criteria:[54]

> ...*whether the action threatens a violation of Federal, State, or local law or requirements imposed for the protection of the environment.*

This regulatory provision could have more simply been written: "... whether the action *violates* a federal...," but instead the word "threatens" was used. This wording appears to imply that the regulations' author(s) believed that if an impact *approached* a regulatory threshold, the impact could be regarded as being significant even if it did not actually breach or violate the threshold. In other words, the act of "significantly" increasing an environmental baseline—making it that much easier for an actual violation to occur sometime in the future—might be viewed as a significant impact. The increased baseline noise level increases the likelihood that future noise sources could push the cumulative noise levels to objectionable levels.

Case 4: A small incremental impact is sufficient to breach the significance threshold Illustrated by Figure 1.7 is a final case, where the cumulative impact baseline noise level is 33 dB, slightly below the threshold of significance (35 dB). In this case, however, the proposed project would increase the cumulative impact baseline from 33 to 35 dB.

While the increase in noise level is relatively small, it is nevertheless sufficient to breach the significance threshold of 35 dB. Because the impact of this proposal would breach the significance threshold, it constitutes a significant cumulative impact. In this case, preparation of an EIS is warranted because alternatives or mitigation measures might be identified that can reduce the impact enough to prevent the significance threshold from being breached.

Factors used in assessing significance

In conformance with a strict or more standard interpretation, both the context and intensity factors described in §1508.27 must be considered in reaching an SDP determination of significance. As illustrated by the four cases above, one of the CEQ's ten significance factors (i.e., §1508.27 [b][2]) could be assessed in terms of the *degree to which* the project would cause a *change* in existing public health and safety (i.e., noise). An action's effect on public health and safety could be considered nonsignificant as long as the cumulative impact on public health and safety does not substantially change or depart from conditions that would exist if the action were not pursued.

Specific guidelines could even be issued to assist decision makers in determining how the ten significance factors should be interpreted, in terms of the SDP. Specific criteria might also be developed to assist agencies in determining when an impact causes a *significant departure* in the environmental impact baseline. For instance, the severity or magnitude of a cumulative impact on a particular species might depend on the amount of habitat that has been disrupted or will be disrupted in the future. Similarly, the magnitude of a cumulative effect on archaeological resources might be quantified or measured by estimating the number of past, present, and future sites or artifacts that have been or will be disrupted. Likewise, the threshold of cumulative significance for a visual impact might be measured based on the perceptions (e.g., low, medium, high) of visitors to a recreational area that is the site for a proposed action. In many cases, computer models can be applied to predict cumulative changes such as socioeconomic patterns, financial conditions, or hydrological conditions.

Fred March summarizes seven specific tests for "significance"[55] that are based on the significance factors defined in the CEQ's NEPA regulation.[56] Based on March's scheme, the SDP does not suggest an eighth independent test; instead, it can be applied in assessing and interpreting some of his seven tests.

Preparing NEPA programmatic analyses and tiering

Programmatic or strategic EISs and EAs are prepared for proposed policies, plans, programs, or projects that are broad in their reach and will later be implemented through site-specific projects that may be analyzed in subsequent, tiered, NEPA analyses. To this end, the following section provides direction for preparing broad programmatic environmental impact statements (PEISs) and programmatic environmental assessments (PEAs) that will support tiering of later project-specific EISs and EAs.

According to the CEQ, a PEA or PEIS provides one of the most effective tools for assessing cumulative impacts. This section was prepared to provide practitioners with guidance on preparing programmatic analyses that can be used to effectively shape broad policies, programs, plans, with emphasis on assessing, and mitigating the resulting cumulative impacts.

Regulatory guidance for preparing programmatic analyses

If a proposed program is under review, it is possible that site-specific actions are not yet proposed. In such a case, these actions are not addressed in the EIS on the program, but are reserved for a later tier of analysis. The term "programmatic" refers to broadly scoped analyses that assess the environmental impacts of federal actions across a span of conditions, such as facilities, geographic regions, or multi-project programs.

While the NEPA regulations (regulations) do not specifically define "broad actions," they do state that (40 CFR §1502.4[b)]):

> *Environmental impact statements may be prepared, and are sometimes required, for broad Federal actions such as the adoption of new agency programs or regulations. Agencies shall prepare statements on broad actions so that they are relevant to policy and are timed to coincide with meaningful points in agency planning and decision-making.*

A programmatic approach typically limits the depth of analyses to the broad impacts of a broad proposal, e.g., the potential nation or region wide impacts of a new proposed federal policy, plan, program, or project. The regulations go on to identify three approaches agencies can use in preparing statements on broad actions:

> (1) *Geographically, including actions occurring in the same general location, such as body of water, region, or metropolitan area.*
>
> (2) *Generically, including actions that have relevant similarities, such as common timing, impacts, alternatives, methods of implementation, media, or subject matter.*
>
> (3) *By stage of technological development, including federal or federally assisted research, development, or demonstration programs for new technologies which, if implemented, could significantly affect the quality of the human environment.*

Tiering is a method "to relate broad and narrow actions and to avoid duplication and delay." Specifically, it can be used to support development of a subsequent project- or site-specific projects assessment from a broader policy, plan, program, or project analyses (40 CFR §1502.20):

> *Agencies are encouraged to tier their environmental impact statements to eliminate repetitive discussions of the same issues and to focus on the actual issues ripe for decision at each level of environmental review (Section 1508.28). Whenever a broad environmental impact statement has been prepared (such as a program or policy statement) and a subsequent statement or environmental assessment is then prepared on an action included within the entire program or policy (such as a site specific action) the subsequent statement or environmental assessment need only summarize the issues discussed in the broader statement and incorporate discussions from the broader statement by reference and shall concentrate on the issues specific to the subsequent action. The subsequent document shall state where the earlier document is available. Tiering may also be appropriate for different stages of actions. (Section 1508.28).*

Tiering is defined as (40 CFR §1508.28):

> *Tiering refers to the coverage of general matters in broader environmental impact statements (such as national*

*program or policy statements) with subsequent narrower
statements or environmental analyses (such as regional
or basin-wide program statements or ultimately site-specific statements) incorporating by reference the general
discussions and concentrating solely on the issues specific
to the statement subsequently prepared. Tiering is appropriate when the sequences of statements or analyses are:*

(a) *From a program, plan, or policy environmental
impact statement to a program, plan, or policy
statement or analysis of lesser scope or to a site-
specific statement or analysis.*

(b) *From an environmental impact statement on a
specific action at an early stage (such as need and
site selection) to a supplement (which is preferred)
or a subsequent statement or analysis at a later
stage (such as environmental mitigation). Tiering
in such cases is appropriate when it helps the lead
agency to focus on the issues which are ripe for
decision and exclude from consideration issues
already decided or not yet ripe.*

Programmatic analyses

A well-designed programmatic analysis can provide the basis for broad
preliminary decisions, such as identifying geographically bounded
areas within which future activities will be designated. Such an analysis
can also facilitate an early understanding of programmatic cumulative
impacts that the subsequent, more specific analysis can tier from. A programmatic or strategic impact analysis can also shape subsequent sustainability planning decisions, and thereby support a sound, integrated,
and sustainable policy or development plan.

Another important benefit of programmatic analyses is that they are
useful in supporting policy- and planning-level decisions when available information is limited or there is uncertainty regarding the timing,
location, and site-specific impacts of a subsequent action. For example,
although an agency may not be able to predict with certainty the environmental impacts of future project-specific project, they may still be able to
make broad program decisions based on a programmatic understanding
of impacts.

Programmatic NEPA approaches

One approach relies on a programmatic analysis to analyze the impacts
and alternatives of a broad policy, program, plan, or project, and then use
tiered analyses to adequately investigate the site-specific issues based on

an understanding of the alternatives and impacts investigated in the programmatic analysis. (See comparison in Appendix B.) Under this approach,
the agency should consider the scope of the programmatic NEPA analysis.
Guidance issued by the CEQ in 1983 states that[57]

> *Tiering, of course, is by no means the best way to handle
> all proposals which are subject to NEPA analysis and
> documentation. The regulations do not require tiering;
> rather, they authorize its use when an agency determines
> it is appropriate. It is an option for an agency to use when
> the nature of the proposal lends itself to tiered EIS(s).*
>
> *Tiering does not add an additional legal require-
> ment to the NEPA process. ... tiering has been estab-
> lished as an option for the agency to use, as opposed to
> a requirement. ... the Council believes that tiering can
> be a useful method of reducing paperwork and duplica-
> tion when used carefully for appropriate types of plans,
> programs and policies which will later be translated into
> site-specific projects. Tiering should not be viewed as an
> additional substantive requirement, but rather a means
> of accomplishing the NEPA requirements in an efficient
> manner as possible.*

Programmatic analyses prepared when the agency is not making
decisions are not NEPA programmatic analyses. Previous CEQ NEPA
guidance states that[58]

> *In geographic settings where several Federal actions are
> likely, to have effects on the same environmental resources
> it may be advisable for the lead Federal agencies to coop-
> erate to provide historical or other baseline information
> relating to the resources. This can be done either through
> a programmatic NEPA analysis or can be done separately,
> such as through a joint inventory or planning study. The
> results can then be incorporated by reference into NEPA
> documents prepared for specific Federal actions so long as
> the programmatic analysis or study is reasonably avail-
> able to the interested public.*

The aforementioned CEQ guidance and regulatory direction has
largely been upheld by the courts. For example, this direction is reinforced by a court case; specifically, a programmatic analysis reflects the
broad environmental consequences attendant upon a wide-ranging federal program by focusing on broad issues relevant to the program while

a subsequent site-specific assessment will address more particularized considerations.[59]

In another case, a court concluded with this direction; a programmatic analysis considers the environmental consequences of a project as a whole, and does not necessarily contain the same level of detail or specificity as a site- or project-specific EIS. The programmatic analysis provides a basis for tiering future NEPA documents focusing on specific facets under review.[60]

According to another court, a programmatic analysis would not require "…consideration of detailed alternatives with respect to each aspect of the plan—otherwise a programmatic analysis would be impossible to prepare and would merely be a vast series of site specific analyses."[61]

Policies, programs, plans, and area-wide analyses

Programmatic NEPA analyses tend to fall into the principal major categories shown below. For proposed actions falling within any of these categories, agencies may use a phased decision-making strategy. That is, agencies may prepare broad programmatic NEPA analyses from which they tier subsequent, more detailed, project-specific analyses for specific proposals implementing the program.

1. *Adopting an Official Policy:* Decisions involving adoption of an official policy in a formal document establishing an agency's policies that will result in or substantially alter agency programs. These programmatic analyses should include a road map for future agency actions with defined objectives, priorities, rules, and mechanisms to implement objectives. Specific examples include
   a. Adoption of an agency-wide policy
   b. National-level rulemaking
2. *Adopting a Formal Plan:* A decision to adopt a formal plan, such as documents that guide or prescribe alternative uses of federal resources, upon which future agency actions will be based (setting priorities, options, and measures for future resource allocation according to resource suitability and availability). Specific examples include:
   a. Adoption of an agency plan for a group of related projects
   b. Strategic planning linked to agency resource allocation
3. *Adopting an Agency Program:* A decision to proceed with a group of concerted actions to implement a policy or plan; an agenda with defined objectives to be achieved during program implementation, with specification of activities. Specific examples include
   a. Proposals to substantially redesign existing programs
   b. A new agency mission or initiative
4. *Approving a Site-Wide or Area-Wide Actions:* A decision to proceed with multiple projects that are temporally or spatially connected

Table 1.7 General Differences Between Programmatic and
Subsequent (Often Project-Level) Tiered Analyses

	Programmatic level	Subsequent or site-specific tiered level
Nature of action	Strategic, conceptual	Construction, operations, site-specific actions
Level of decision	Policy, program, planning	Projects
Alternatives	Broad, general, research, technologies, fiscal measures, economic, social, regulatory	Specific alternative locations, design, construction, operation, permits, site specific
Scale of impacts	Macroscopic, for example, at a national regional, or landscape level	Project level, mainly local
Scope of impacts	Broad in scale and magnitude; particularly useful for assessing cumulative impacts	Localized and specific
Time scale	Long- to medium-term (Regulatory)	Medium- to short-term (Permit)
Key data sources	Existing national or regional statistical and trend data, policy and planning instruments	Field work, sample analysis, statistical data, local monitoring data
Impacts	Qualitative and may be quantitative to the degree possible	Quantifiable
Decision	Broad, strategic program, policy, or plan	Detailed, site-specific, action-oriented

and have a series of associated subsequent or concurrent decisions.
Specific examples include

a. A suite of ongoing, proposed, or reasonably foreseeable actions
   that share a common geography or timing, such as multiple
   functions within the boundaries of a large federal facility
b. Several similar actions or projects in a region or nationwide

Differences between programmatic and tiered analyses

Programmatic and tiered analyses differ in their focus and scope. Table 1.7
delineates some of general differences between programmatic and subsequent, often project-level, tiered analyses.[62]

Programmatic analysis guidance

The scoping process provides a means to develop the scope of the analysis and assist the agency in addressing these attributes described in
Table 1.7. A programmatic analysis can become overly "bloated," costly,

and may provide little in terms of usefulness if it is not properly scoped at the early proposal stage. The companion book entitled *NEPA and Environmental Planning* provides a tool consisting of eight criteria for assisting practitioners and decision makers in determining the appropriate scope of the programmatic analysis. The scope of the analysis needs to address[63]

- Three types of actions: connected, cumulative, and similar
- Three types of alternatives: no action, other reasonable courses of actions and mitigation
- Three types of impacts: direct, indirect, and cumulative

The scoping process should include interested stakeholders (federal, state, tribal, and others) as early in the NEPA process as possible. An earnest engaging of stakeholders during the early planning stage also provides an opportunity to develop trust and good working relationships that may extend throughout the planning and implementation process.

Actions   Connected and cumulative actions should be included in a programmatic NEPA analysis and document, and the responsible official must consider whether to also include any similar actions (40 CFR §1508.25(a)).

Alternatives   Programmatic alternatives should provide a well-defined scope for the next level of decision making. This allows agencies to develop focused alternatives in the programmatic document, which limits the scope and alternative development of the subsequent tiered NEPA document.

With respect to alternatives, one court had this to say: "[t]he critical inquiry in considering the adequacy of an EIS prepared for a large scale, multi-step project is the project's site-specific impact should be evaluated in detail, but when such detailed evaluation should occur."(40 CFR §1501.7). The court added that this "threshold is reached when, as a practical matter, the agency proposes to make an "irreversible and irretrievable commitment of the availability of resources" to a project at a particular site.[64] Another court provided this direction: "an EIS for a programmatic plan must provide sufficient detail to foster informed decision making, but that site-specific impacts need not be fully evaluated until a critical decision has been made to act on site development."[65]

Impacts   Because the impacts covered in a programmatic analysis typically range over a broad geographic and time horizon, the depth and detail of impact analysis is expected to be broad and general for programmatic NEPA analyses and documents. The effects analyses will focus on the major impacts that might result from implementing a broad federal action and, especially on those resources or factors that are adversely impacted. Such impacts are typically evaluated in a broad geographic and temporal context with particular emphasis on cumulative impacts. The scope and range of impacts may be more qualitative in nature than those found in project NEPA documentation. The following questions may be helpful when considering the scope of impacts to be considered in a programmatic NEPA analysis and document:

1. What environmental, social, or economic impact criteria are of most concern at this programmatic scale?
2. For each potentially affected environmental resource, what is the appropriate geographic scale of the affected environment (e.g., basin, watershed, etc.) for the programmatic setting?

Addressing decisions and issues in tiered analyses

The relationship between the programmatic document and future or subsequent tiered documents should be described in the programmatic document. Decisions or analyses that are deferred to future documents should be articulated in the programmatic analysis.[66]

The programmatic document should also explain how and when the interested parties will be notified of any subsequent analyses identified in the programmatic NEPA document. Examples successfully used by several agencies demonstrating how differed decision can be implemented in tiered or later analyses are depicted in Table 1.8.

New information

The CEQ NEPA Regulations specify that implementation of the action should be accompanied by monitoring in important cases and provide a procedural framework for keeping environmental analyses current by requiring that agencies prepare supplements if significant new information of relevance to the proposed action or its impacts are discovered.

The possibility of new information arising after a final EA or an EIS is completed exists whether or not that NEPA document is a programmatic one. However, some agencies are more concerned about the possible effects of the receipt of new information on programmatic NEPA documents because they normally have a broader scope and a longer projected life span.

If new information is uncovered following a programmatic analysis, it should be evaluated with respect to the following considerations:

1. Does the new information pertain to a programmatic NEPA document that was prepared for a now-completed decision-making process? Phrased in the alternative, are there any more decisions to be made by the agency that would use the original NEPA document to meet all or a portion of the agency's NEPA compliance responsibilities for any upcoming decision?

Table 1.8 Examples of Programmatic Analyses

Example of broad or programmatic analysis	Why analysis was used	Trigger for further analysis or action	How stakeholders become aware of further analysis or actions
Agency policymaking Ex: USDA Fruit Fly Cooperative Control Program Final Environmental impact statement (EIS)—2001	Introduction of invasive fruit fly species can occur at multiple potential sites throughout the US. The EIS evaluates broad issues such as potential locations, control strategies, mitigation measures, and cumulative impacts avoids segmentation of analyses and provides basic information to foster efficiency by focusing the scope on critical issues that will be analyzed for site-specific assessments.	The detection of a non-native, invasive fruit fly species introduction at levels determined to be sufficient for establishment is the trigger for agency action and the preparation of a site-specific EA tiered to the EIS.	Each site-specific EA has its own public involvement process with associated public comment period.
Geographic or regional action Ex: DOT "Transportation Corridor" Tier I EIS	The EIS examines broad issues such as general location, mode choice, air quality, and land use implications of major alternatives	As site-specific projects are identified, each project will have a separate Tier II EA/EIS. Tier I EIS specifies decisions that must be resolved in Tier II documents.	Each site-specific Tier II project will have its own public involvement process, as specified in the Tier I EIS and ROD.

(Continued)

Table 1.8 Examples of Programmatic Analyses (Continued)

Example of broad or programmatic analysis	Why analysis was used	Trigger for further analysis or action	How stakeholders become aware of further analysis or actions
Technical program with a combination of known elements or conditions Ex: NASA's environmental assessment for routine payloads on expendable launch vehicles	Analyzed common launch vehicles, two common launch sites, and broad classes of payload risk. Allowed short-turnaround of projects within known risks.	Each new project completes a checklist to identify launch vehicle, launch site, and payload. Any of these parameters outside of those listed in the EA would result in a supplemental analysis (e.g. project EA).	Supplemental analyses (where required) are publicly announced in a manner similar to the original Programmatic EA (regional newspapers, local public meetings, etc.).
Range of activities and operations within a facility Ex: Department of Energy (DOE) Programmatic Spent Nuclear Fuel Management and Idaho National Engineering Laboratory Environmental Restoration and Waste Management Programs (DOE/EIS-0203, April 1995)	The EIS supports two sets of decisions: (1) DOE-wide decisions on spent nuclear fuel (SNF) management (Volume 1), and (2) site-specific decisions on the future direction of environmental and waste management programs at the Idaho National Engineering Laboratory (now called the	In the analysis of broad program alternatives, the PEIS considered the individual and collective environmental impacts of ongoing activities at INL and also reasonably foreseeable future projects. As explained in the PEIS and ROD, the PEIS was intended to support	If DOE proposes to implement a specific project, additional NEPA review (e.g., an EA or EIS) would be conducted, with appropriate further public participation. DOE has completed several such tiered EISs under this PEIS.

Table 1.8 Examples of Programmatic Analyses (Continued)

Example of broad or programmatic analysis	Why analysis was used	Trigger for further analysis or action	How stakeholders become aware of further analysis or actions
		implementation decisions for a defined set of proposed projects; other projects were analyzed to ensure adequate cumulative impacts analysis. The "trigger" for further analysis would be a DOE proposal to implement one of the other specific projects.	

If no further decisions need to be made, there is no need to supplement the original programmatic analysis, with respect to the issue of new information. This result could occur for a programmatic NEPA document completed for a one-time decision. However, if

2. The new information is relevant to a future decision for which the agency intends to rely upon the original programmatic NEPA document to meet all or a portion of its NEPA compliance responsibilities, then a second consideration must be addressed. The new information must be reviewed in order to determine if it has any potential effect on the content of the original programmatic document, either in terms of (a) the accuracy of the previously analyzed impacts (direct, indirect or cumulative) or (b) the feasibility of the alternatives presented or their comparative analysis.

If the response to this question is "No," then further agency action is not required with respect to the original programmatic document.

If the original analysis was a PEA, a similar determination should be performed. The focus of this determination, however, must include whether or not an EA and its FONSI still suffice or whether an EIS is now

necessary. The PEA should be supplemented to address the new information and the FONSI reconsidered based upon the agency's significant
impact criteria, which would include consideration of the context and
intensity of the effects of the programmatic action (40 CFR §1508.27).

Litigation and judicial review

Historically, federal agencies have had a poor track record in defending
their NEPA documents against legal challenges based on inadequate CIAs.
Over the past decade, litigation has increasingly focused on cumulative
impacts. The number of cases where the courts have found the analyses
of cumulative impacts to be inadequate has been proportionately higher
than for those cases involving the analysis of direct impacts only.

Smith examined 25 recent judicial opinions from the Ninth Circuit
Court of Appeals involving such challenges.[67] With respect to challenges
based on an inadequate CIA, plaintiffs were successful in 60% of the cases
over a 10-year period. In recent years, the success rate of plaintiffs has
risen even further, to victories in 72% of the cases (8 out of 11 verdicts).
Based on his study, the federal agency with the worst record was the US
Bureau of Land Management, which lost all three of its cases (100%); this
was followed by the US Forest Service, which lost 69% (9 of 13) of its cases,
and the US Army Corps of Engineers which lost 66 percent (two of three
cases) of the time. The principal reasons that federal agencies lost these
court cases were because they

- Left out obvious past, present, or reasonably foreseeable future actions
- Presented unfounded assertions that their projects would not cause
  any significant cumulative impacts
- Failed to have any assessment of cumulative impacts whatsoever in
  their NEPA document

Examples of court direction

In performing a CIA, an agency is responsible for providing some degree
of quantified or detailed information; general statements about possible
effects or some risks do not constitute a hard look absent a justification
regarding why more definitive information could not be provided.[68] More
to the point, a cumulative analysis must be more than perfunctory; as one
court stated, it must provide "a useful analysis of the cumulative impacts
of past, present, and future projects."[69]

The US Supreme Court ruled that whenever an EIS is prepared, it
should include an analysis of other activities either related to or similar
to the proposed action.[70] This concept paved the way for a later requirement to consider those cumulative impacts that could result from past,

present, and reasonably foreseeable future actions. The CEQ, recognizing the importance of considering impacts in a cumulative sense, codified this requirement in the regulations.

In a second example concerning cumulative impacts, the Sierra Club challenged an EA prepared by the DOE for the shipment of spent nuclear fuel through the Port of Hampton Roads, Virginia.[71] Under the proposal, the fuel was to be transported in not one but several successive shipments, thereby increasing the risks to this population of a possible accident and its resulting impacts. Yet, the analysis did not consider the *cumulative* potential risk of accidents and impacts from permitting several shipments. Because of its failure to make this assessment, the court found the analysis to be flawed.

Court direction on performing cumulative impact assessments

The US Supreme Court has stated that agencies may properly limit the scope of their cumulative effects analysis based on practical considerations.[72] As indicated in Table 1.9, the Supreme Court has also provided five criteria that a meaningful cumulative impacts analysis needs to meet.[73]

This Supreme Court direction does not necessarily imply that an agency must prepare a "full-blown" analysis of all the actions considered in a cumulative analysis, equivalent to the type of analysis performed in other reviews. Instead, the Supreme Court stated that other actions and their probable impacts had to be *identified* and *considered* in determining whether the proposal could result in a significant impact.[74]

In a second case, the Federal Aviation Administration (FAA) was challenged for addressing only the incremental increase in noise and not the cumulative impacts that could result from replacing an airport near Zion National Park. There was no way to determine from the proposal whether the FAA's estimated 2% increase would actually result in a significant impact if it was *added to* other existing noise impacts on the park.[75]

In a third case, a plaintiff sued, claiming that an EIS prepared for a post-fire logging project in one section of a forest did not adequately disclose or analyze its potential cumulative impacts on the management

Table 1.9 Supreme Court Direction on Performing a Cumulative Impact Analysis

- Specify the area that would be affected by the proposed action
- Identify the potential impacts expected to occur within the affected area
- Address other past, proposed, and reasonably foreseeable actions that have affected or might affect the area
- Evaluate the identified impacts that have resulted or are expected to result from these actions
- Describe the cumulative impacts that can be expected if these individual impacts are allowed to accumulate

of indicator species, fuel break maintenance, or fire-fighting tactics when added to the impacts of other post-fire logging projects in that forest. The court concluded that given the similarities of these projects with respect to timing, geography, and purpose, such actions might result in a significant *cumulative* impact that needed to be addressed in a single EIS.[76]

Reasonably foreseeable actions   The Supreme Court has recognized the importance of considering "reasonably foreseeable actions" in terms of those needing identification and evaluation as part of the CIA. Here, the Supreme Court provided a broad interpretation of the term "reasonably foreseeable" to include even the future impacts of projects that had not yet been formalized.[77]

In one case, a US Forest Service proposed to construct a road through a roadless forest.[78] In response, a suit was brought, stating that the agency needed to evaluate the *cumulative* impacts to the environment resulting from the logging and timber sales that would be triggered by the construction. The Forest Service argued that because no sales would be contemplated for many years into the future, such sales were too uncertain to be evaluated as a part of the road building project. The court reasoned that if this were the case, their argument was tantamount to admitting that constructing the road was senseless. It went on to conclude that the Forest Service proposal not only precluded the analysis of other reasonable alternatives, but also swayed the final decision concerning future land use in favor of timber sales. The court held that if the cumulative impacts from both the road and the potential timber sales had been analyzed together, the agency might have reached a different conclusion. In the words of the court, "[If] sales are sufficiently certain to justify construction of the road, then they are sufficiently certain for their environmental impacts to be analyzed along with the road."

Differences in cumulative impact analyses between EAs and EISs

One theme emphasized throughout this book is that while EAs are generally shorter documents than EISs, they are not always simpler documents nor are they always easier to prepare. The same theme is true when contrasting the effort needed to successfully perform a CIA in an EA versus an EIS. As noted in other chapters, agencies relying on an EA and FONSI in lieu of an EIS bear the burden of proof in demonstrating that no significant impact—direct, indirect, or cumulative—would result from a proposed action. That burden of proof is alleviated should an EIS be prepared. An EIS needs only to present an assessment of potential environmental impacts, not prove that those impacts are not significant. In determining whether an EIS needs to be prepared, the analysis of cumulative effects in an EA may sometimes actually require more rigor than needed in an EIS.

The Fifth Circuit Court decision   The court concluded that the question of determining when an EIS is required may necessitate a broader analysis of cumulative impacts than is generally necessary in an EIS. According to the court, an EA "should consider (1) past and present actions without regard to whether they themselves triggered NEPA responsibilities, and (2) future actions that are reasonably foreseeable, even if they are not yet proposals and may never trigger NEPA-review requirements."[79]

Specifically, the Fifth Circuit found that

> *...although cumulative impacts may sometimes demand*
> *the preparation of a comprehensive EIS, only the impacts*
> *of proposed, as distinguished from contemplated, actions*
> *need be considered in scoping an EIS. In a case like this*
> *one, on the other hand, where an EA constitutes the only*
> *environmental review undertaken thus far, the cumula-*
> *tive impacts analysis plays a different role...*
>
> *[The NEPA Regulations] require an analysis, when*
> *making the NEPA-threshold decision, [to determine if]*
> *it is reasonable to anticipate cumulatively significant*
> *impacts from the specific impacts of the proposed project...*
> *[W]hen deciding the potential significance of a single pro-*
> *posed action ... a broader analysis of cumulative impacts*
> *is required. The regulations clearly mandate consideration*
> *of the impacts from actions that are not yet proposals and*
> *from actions—past, present, or future—that are not them-*
> *selves subject to the requirements of NEPA.*

The court cautioned that it did not mean to imply that "...consideration of cumulative impacts at the threshold stage will necessarily involve extensive study or analysis of the impacts of other actions." Instead, the court emphasized that the CIA in an EA should be limited to determining whether "...the specific proposal under consideration may have a significant impact."

The court went on to state that, at a minimum, an EA must demonstrate that the agency considered impacts from "past, present, and reasonably foreseeable future actions regardless of what agency, (federal or non-federal), or person undertakes such other actions." According to the court, the extent of the analysis depends on the scope of the affected area and the extent of other past, present, and future activities.

Endnotes

1. CEQ (Council for Environmental Quality), Considering Cumulative Effects under the National Environmental Policy Act, p. 1, January 1997.
2. CEQ (Council for Environmental Quality), Considering Cumulative Effects under the National Environmental Policy Act, at vi, January 1997.
3. CEQ (Council for Environmental Quality), Considering Cumulative Effects under the National Environmental Policy Act, p. 4, January 1997.
4. CEQ (Council for Environmental Quality), Considering Cumulative Effects under the National Environmental Policy Act, p. 128, January 1997.
5. CEQ (Council for Environmental Quality), Considering Cumulative Effects under the National Environmental Policy Act, p. 4, January 1997.
6. Eccleston, C. H., *Environmental Impact Statements: A Comprehensive Guide to Project and Strategic Planning,* New York: John Wiley & Sons, 2000.
7. Final Environmental Assessment and Finding of No Significant Impact, US Government Easements and Water Reallocation at John H. Kerr Reservoir, Mecklenburg Cogeneration Plant, Mecklenburg County, Virginia. US Army Corps of Engineers, Wilmington District, Wilmington, NC, March 1991.
8. Cumulative Effects Working Group et al., 1999.
9. Kleppe, 427 U.S. at 414.
10. CEQ (Council for Environmental Quality), Considering Cumulative Effects under the National Environmental Policy Act, p. 17, January 1997.
11. CEQ (Council for Environmental Quality), Considering Cumulative Effects under the National Environmental Policy Act, p. 17, January 1997 (figure is after Bisset 1983).
12. CEQ (Council for Environmental Quality), Considering Cumulative Effects under the National Environmental Policy Act, p. 15, January 1997.
13. *Northern Alaska Environmental Center v. Lujan,* 15 ELR 21048 (D. Alaska, 1985).
14. *Sierra Club v. Forest Service,* 46 F.3d 835, 839 (8th Cir. 1995).
15. *Sierra Club v. U.S. Forest Service* (9th Cir. June 24, 1988).
16. *Save the Yaak Committee v. Block,* 840 F.2d 714 (9th Cir. 1988).
17. *Thomas v. Peterson,* 753 F.2d 754 (9th Cir. 1985).
18. CEQ, Considering Cumulative Effects under the National Environmental Policy Act, p. 19, January 1997.
19. *Webb v. Gorsuch,* 699 F.2d 157, 161 (4th Cir. 1983).
20. *Fritiofson v. Alexander,* 722 F.2d 1225 (5th Cir. 1985).
21. 379 F.3d 738 (9th Cir. 2004); amended 395 F.3d 1019 (9th Cir. 2005).
22. CEQ (Council for Environmental Quality), Guidance on the Consideration of Past Actions in Cumulative Effects Analysis, June 24, 2005, http://ceq.eh.doe.gov/nepa/regs/Guidance_on_CE.pdf.
23. No.CV-05-0220-EFS, 2005 WL 2077807 (E.D. Wash., Aug. 26, 2005) (citing *Andrus v. Sierra Club,* 442 U.S. 347, 358 [1979]).
24. *Conservation Northwest v. USFS,* No.CV-05-0220-EFS, 2005 WL 2077807, (E. D. Wash. Aug. 26, 2005) (citing *Andrus v. Sierra Club,* 442 U.S. 347, 358 [1979]); *Native Ecosystem Council v. U.S. Forest Service,* 428 F.3d 1233 (9th Cir. 2005); *Northwest Environmental Advocates, et al. v. National Marine Fisheries Service, et al.,* 460 F.3d 1125 (9th Cir. 2006); *Environmental Protection Information Center v. U.S. Forest Service,* 451 F.3d 1005 (9th Cir. 2006).

25. Magee, J., and Nesbit, R., Proximate causation and the no action alternative trajectory in cumulative effects analysis, *Environmental Practice*, 10(3), 110, 2008.
26. *Department of Transportation v. Public Citizen*, 541 U.S. (2004).
27. 460 U.S. 766, 774 (1983).
28. *Department of Transportation v. Public Citizen*, 541 U.S. at 767.
29. 460 U.S. 766, 774 (1983).
30. 541 U.S. 752 (2004).
31. *Department of Transportation. v. Public Citizen*, 541 U.S. at 767.
32. *Department of Transportation v. Public Citizen*, 541 U.S. at 769–70.
33. Keeton, W. P., Dobbs, D. B., Keeton, R. E., Owen, D. G., and W. L. Prosser, W. L., Editors, *Prosser and Keeton on the Law of Torts*, 5th edition, §41, at 264, St. Paul, MN: West Publishing, 1984.
34. Magee J., and Nesbit, R., Proximate causation and the no action alternative trajectory in cumulative effects analysis, *Environmental Practice*, 10(3), 107–115, 2008.
35. Eccleston, C. H., *NEPA and Environmental Planning: Tools, Techniques, and Approaches for Practitioners*, Boca Raton, FL: CRC Press, pp. 236–237, 2008.
36. CEQ (Council for Environmental Quality), Considering Cumulative Effects under the National Environmental Policy Act, p. 17, January 1997, at v, 1997.
37. *Fritiofson v. Alexander*, 772 F.2d 1225, 1243, 1245–6 (5th Cir. 1985).
38. *Inland Empire Public Lands Council v. Schultz*, 992 F.2d 977, 982 (9th Cir. 1993).
39. CEQ, Considering Cumulative Effects under the National Environmental Policy Act, p. 17, January 1997, January 1997.
40. 40 CFR §1508.27 (b)(7).
41. Eccleston, C. H., Applying the significant departure principle in resolving the cumulative impact paradox, *Environmental Practice*, 8(4), 241–250, December 2006.
42. CEQ (Council for Environmental Quality), The National Environmental Policy Act: A Study of Its Effectiveness after Twenty-Five Years, p. 19, January 1997.
43. 40 CFR §1508.13.
44. 40 CFR §1508.4.
45. 40 CFR §1500.4, Reducing Paperwork; 40 CFR §1500.5, Reducing Delay.
46. 40 CFR §1508.28(a).
47. 40 CFR §1508.27, Significantly.
48. 40 CFR §1508.27 (b)(2).
49. 40 CFR §1508.27 (b), Significantly.
50. 40 CFR §1508.7.
51. CEQ (Council for Environmental Quality), Considering Cumulative Effects under the National Environmental Policy Act, January 1997.
52. CEQ (Council for Environmental Quality), Guidance on the Consideration of Past Actions in Cumulative Effects Analysis, June 24, 2005, http://ceq.eh.doe.gov/nepa/regs/Guidance_on_CE.pdf.
53. 40 CFR §1508.27.
54. 40 CFR §1508.27(b)(10).
55. Eccleston, C. H., *Effective Environmental Assessments: How to Manage and Prepare NEPA EAs*, Lewis Publishers, Chapter 8, 2001. Reference to Fred March.

56. 40 CFR §1508.27.
57. 48 Fed. Reg. 34263 (1983): CEQ 1983 Guidance Regarding NEPA Regulations.
58. CEQ (Council on Environmental Quality), Guidance on the Consideration of Past Actions in Cumulative Effects Analysis, June 24, 2005, available at http://ceq.eh.doe.gov/nepa/regs/Guidance_on_CE.pdf.
59. *Nevada v. Dep't of Energy*, 372 U.S. App. D.C. 432 (D.C. Cir. 2006).
60. *Greenpeace v. National Marine Fisheries Service*, 55 F. Supp. 2d 1248, 1273 (D. Wash. 1999).
61. *Greenpeace v. National Marine Fisheries Service*, 55 F. Supp. 2d 1248, 1276 (D. Wash. 1999).
62. CEQ (Council for Environmental Quality), Draft NEPA Programmatic Guidance, September 18, 2007. Table based on *Strategic Environmental Assessment (SEA)—Current Practices, Future Demands and Capacity-Building Needs*, a course manual by Maria Rosário Partidário, International Association for Impact Assessment Training, 2003.
63. Eccleston, C. H., *NEPA and Environmental Planning: Tools, Techniques, and Approaches for Practitioners*, Boca Raton, FL: CRC Press, pp. 223–227. 2009.
64. *California v. Block*, 690 F.2d 753, 761, (9th Cir. 1982).
65. *Citizens for Better Forestry v. United States Department of Agriculture*, 481 F. Supp. 2d 1059, 1086, (D. Cal. 2007).
66. CEQ (Council for Environmental Quality), Draft NEPA Programmatic Guidance, September 18, 2007.
67. Smith, M. D., National Association of Environmental Practice, "An Analysis of Recent NEPA Cumulative Impact Assessment Case Law, *Environmental Practice*, 8(4), 241–250, December 2006.
68. *Neighbors of Cuddy Mountain*, 137 F.3d at 1379–80.
69. Kern, 284 F.3d at 1075 (quoting Muckleshoot Indian Tribe (quoting *Muckleshoot Indian Tribe v. United States Forest Service*, 177 F.3d 800, 810 [9th Cir. 1999]).
70. *Kleepe v. Sierra Club*, 427 U.S. 390, 1976.
71. *Sierra Club v. Watkins*, 808 F. Supp. 852, 1991.
72. *Kleppe*, 427 U.S. at 414.
73. *Fritiofson v. Alexander* (CA5, 1985) 772 F 2d 1225.
74. *Fritiofson v. Alexander* (CA5, 1985) 772 F 2d 1225.
75. *Grand Canyon Trust v. Federal Aviation Administration*, 290 F.3d 339 (D.C. Cir. 2002).
76. *Sierra Club v. Bosworth*, 199 F. Supp 2d 971 (N.D. Cal. 2002).
77. *Fritiofson v. Alexander* (CA5, 1985) 772 F 2d 1225.
78. *Thomas v. Peterson*, 753 F.2d 754 (9th Cir. 1985).
79. *Fritiofson v. Alexander*, 772 F.2d 1225, 1243, 1245–1246 (5th Cir. 1985).

chapter two

Preparing greenhouse emission assessments

A synopsis of guidance and best professional practices

> *Congress has a plan to fight global warming. It's passed a law that we can lower the temperature dramatically by simply switching from Fahrenheit to Celsius.*

> **—Anonymous**

Until recently, potential consequences of US governmental activities on climate change have been all but ignored; the impacts of the actions of many other nations have fared little better. With the preponderance of scientific evidence for climate change mounting, the best scientific practices and information need to be made available to ensure that greenhouse gas (GHG) emission assessments are integrated into all aspects of federal planning. GHGs include emissions such as carbon dioxide, methane, nitrous oxide, hydrofluorocarbons, perfluorocarbons, and sulfur hexafluoride. Moreover, a 2007 US Supreme Court ruling held that greenhouse gases meet the Clean Air Act's definition of an "air pollutant" and, as such, can be regulated. This ruling has sweeping implications for industry and government entities.[1]

Research on climate change impacts is an emerging and rapidly evolving area of science. To date, there is no universal consensus regarding the extent to which a National Environmental Policy Act (NEPA) document must consider global climate change impacts. However, the NEPA requires federal agencies to support international cooperation by recognizing.[2]

> *the global character of environmental problems and, where consistent with the foreign policy of the United States, lend appropriate support to initiatives, resolutions, and programs designed to maximize international cooperation in anticipating and preventing a decline in the quality of mankind's world environment.*

Moreover, the NEPA was enacted to, *inter alia,* "promote efforts which will prevent or eliminate damage to the environment and biosphere and stimulate the health and welfare of man."[3] Unfortunately, as this chapter is drafted, there is little general consensus concerning the regulatory and scientific methodology that should be employed in analyzing such impacts. This chapter is designed to

1. Provide a compendium of best professional practices for assessing GHG emissions and climate change impacts in environmental impact assessments (EIA) such as NEPA documents.
2. Assess cumulative GHG emission impacts under NEPA—one of the thorniest issues in American environmental law that poses perhaps the most significant challenge to NEPA's regulatory framework in decades. This chapter provides a peer-reviewed technique, referred to as the Sphinx Solution for resolving this vexing NEPA assessment problem.

Although this chapter focuses on preparing GHG assessments in NEPA documents, this direction is also applicable to most other EIAs. Before presenting guidance on addressing GHG impacts in NEPA documents, we briefly introduce a summary of scientific evidence underlying the subject of climate change.

Brief summary of the science behind greenhouse warming

The *greenhouse effect* refers to a controversial rise in mean global temperature as a result of increasing anthropogenic (human-induced) GHGs that absorb and trap infrared radiation. According to this theory, GHGs trap heat within the Earth's surface-troposphere system. The greenhouse effect was first postulated by the famous physicist Joseph Fourier in 1824. Svante Arrhenius, a chemist, was the first to quantitatively describe the greenhouse effect in a paper published in 1896.

Due to human activities, atmospheric carbon dioxide (CO_2), as measured from ice cores) has increased over the past century from 300 to 380 parts per million (ppm), and the average Earth temperature has increased approximately 0.7°C (or about 1.3°F). According to the "Intergovernmental Panel on Climate Change (IPCC) Fourth Assessment Report," as of 2004, human activities are producing nearly 50 billion tons of GHGs annually (measured in CO_2 equivalency).[4] Ambient concentrations of GHGs do not cause direct adverse health effects (such as respiratory or toxic effects), but public health risks and impacts as a result of elevated atmospheric concentrations of GHGs occur via climate change.[5] While CO_2 is not the most potent GHG, because of the voluminous quantities emitted annually, it is the gas of principal concern.

A natural greenhouse effect has shaped the Earth's climate for several billion years of its history. Without its naturally present greenhouse effect, the Earth's average surface temperature would be 14°C (57°F) and possibly as low as −18°C (−0.5°F). The atmosphere of the planet Venus consists principally of CO_2 with an atmospheric pressure in the range of 90 times greater that that of the Earth. Venus is a planet where the greenhouse effect has run out of control. As a result of its greenhouse effect, the surface temperature of Venus is 460°C (860°F).

The gradual warming of the Earth's lower atmosphere over the past century is widely believed result from an *enhanced greenhouse effect*. The term "enhanced" denotes the belief that anthropogenic GHGs are increasing the Earth's natural greenhouse effect. While this effect is controversial, it is widely believed to result from anthropogenic GHG emissions and large-scale changes in land use (e.g., deforestation). More recently, however, there has been a cooling trend that many cite as evidence against anthropogenic warming. The following sections review the scientific evidence for greenhouse warming.

Current status of the debate

As more data and more reliable climate models have contributed to a better understanding of the Earth's complex climate system, much of the scientific community has expressed increased confidence that anthropogenic GHG emissions are warming the Earth's climate. At the same time, scientific debate actually appears to be diverging rather than converging on this issue.

Intergovernmental panel on climate change

The IPCC was established in 1988 by the World Meteorological Organization and the United Nations Environment Program.[6] Every 5 to 7 years, the IPCC synthesizes the most recent climate science findings and presents its report to the public.

In 2007, the IPCC released its fourth and most recent assessment report, *Climate Change 2007: The Physical Science Basis*. This report was prepared by approximately 600 authors from 40 countries, and was reviewed by over 620 experts and governments. The report describes an extensive peer review of the analyses. The fourth assessment report is arguably the most influential IPCC report because it expressed a higher level of confidence in several key findings, which convinced many policymakers of the need to develop more stringent policies to reduce greenhouse emissions. The section entitled "Summary for Policymakers (SPM)" states

- Warming of the climate system is unequivocal.
- Most of the observed increase in global average temperatures since the mid-20th century is *very likely* due to the observed increase in anthropogenic greenhouse gas concentrations.

Principal conclusions about atmospheric changes   Some of the key conclusions cited in the IPCC report are shown in Table 2.1.

Global climate change   With respect to global changes in the climate, Table 2.2 depicts some of the principal observations presented in the 2007 IPCC report:

Additional sources on climate change

For additional information, the reader is directed to the IPCC Website, http://www.ipcc.ch/ipccreports/assessments-reports.htm.[7] The Website of the US Climate Change Science Program is among the best available sources for up-to-date information on the science of climate change;[8] this program integrates federal research on climate and global change, sponsored by 13 federal agencies, and is overseen by the Office of Science and Technology Policy, Council on Environmental Quality (CEQ), the National Economic Council, and the Office of Management and Budget. Table 2.3 provides some additional guidance and references that may be of use.

On October 5, 2009, President Obama signed Executive Order (E.O.) 13514 (Federal Leadership in Environmental, Energy, and Economic

Table 2.1 A Summary of Key Conclusions Cited in the 2007 Intergovernmental Panel on Climate Change Report

- The principal source of the increase in CO_2 is fossil fuel use; however, land-use changes also add a significant contribution.
- The amount of CO_2 in the atmosphere in 2005 (379 ppm) greatly exceeds the natural range over the past 650,000 years (180 to 300 ppm).
- Carbon dioxide, methane, and nitrous oxide have increased markedly as a result of human activities since 1750 and now far exceed pre-industrial values.
- Nitrous oxide concentrations have risen from a pre-industrial value of 270 parts per billion (ppb) to a 2005 value of 319 ppb. More than a third of this rise is due to human activity, primarily agriculture.
- The primary source of the increase in methane is very likely a combination of human agricultural activities and combustion of fossil-fuels. How much each contributes is not well determined.
- The amount of methane in the atmosphere in 2005 (1,774 ppb) greatly exceeds the natural range over the last 650,000 years (320 to 790 ppb).

Table 2.2 Principal Observations Presented in the 2007
Intergovernmental Panel on Climate Change Report

Key observations

Average Arctic temperatures increased at almost twice the global average rate in
the past 100 years.

Greenhouse gases are likely to have caused more warming than we have
observed if not for the cooling effects of volcanic and human-caused aerosols
(global dimming).

Urban heat island impacts have a negligible influence (less than 0.0006°C per
decade over land and zero over oceans) on these measurements.

Eleven of the 12 years between 1995 and 2006 ranked among the top 12
warmest years in the instrumental record (since 1850, near the end of the Little
Ice Age).

Over the past 100 years, warming has caused about a 0.74°C increase in global
average temperature. This has increased from the 0.6°C in the 100 years prior to
the Third Assessment Report.

Mean Northern Hemisphere temperatures during the second half of the twentieth
century were very likely greater than during any other 50-year period in the past
500 years and likely the highest in at least the past 1,300 years.

Since 1961, the ocean has been absorbing more than 80% of the heat added to the
climate system; ocean temperatures have increased to depths of at least 3,000 m
(9,800 ft).

Hurricanes

It is likely (>50%) that there has been some human contribution in increased
hurricane intensity.

It is likely (>66%) that the planet will witness increased hurricane intensity
during the twentuy-first century.

At present, there is no clear trend in the number of hurricanes.

The observed increase in hurricane intensity is larger than computer models
predict.

Since the 1970s, there has been an increase in hurricane intensity in the North
Atlantic and this increase correlates with increased sea surface temperatures.

Snow, ice, rain, and oceans

Antarctic sea ice showed no significant overall trend, consistent with a lack of
warming in that region.

Oceanic warming causes seawater to expand, contributing to sea-level rising.

Sea level rose by an average rate of about 1.8 mm/year between 1961 and 2003.
Between 1993 and 2003, sea level rose by an average of 3.1 mm/year. Whether
this is a long-term trend or just variability is unclear.

Mountain glaciers and snow cover have declined on average in both hemispheres.

Losses from the land-based ice sheets in Antarctica and Greenland are very likely
(>90%) to have contributed to sea level rise between 1993 and 2003.

Table 2.3 Additional Guidance for Preparing GHG Assessments

- Canadian Environmental Assessment Agency, "Incorporating Climate Change Considerations in Environmental Assessment: General Guidance for Practitioners" (November 2003)
- Guidance issued by Massachusetts Executive Office of Energy and Environmental Affairs and the California Association of Environmental Professionals
- A publication by the US Energy Information Administration, "Documentation for Emissions of GHGs in the United States 2003," provides factors useful in evaluating emissions from smokestacks and other sources
- Guidance designed for use in England and Wales includes protocols produced by Levett-Therivel Sustainability Consultants, "Strategic Environmental Assessment and Climate Change: Guidance for Practitioners" (May 2004)
- The World Resources Institute and the World Business Council for Sustainable Development have also developed guidance (see www.ghgprotocol.org/)

Performance) to establish an integrated governmental strategy to make reduction in GHG emissions a priority for U.S. federal agencies. Among other provisions, E.O. 13514 requires federal agencies to measure, report, and reduce their GHG emissions. It directed the U.S. Department of Energy, in concert with some other federal agencies to develop recommended federal GHG reporting and accounting procedures. A 43-page draft guidance document, "Draft Federal Greenhouse Gas Accounting and Reporting," has been prepared which outlines government-wide requirements for federal agencies in calculating and reporting GHG emissions associated with agency operations. This draft guidance was accompanied by a separate 150-page technical document, "Draft Technical Support Document for Federal GHG Accounting and Reporting (TSD)" which provides detailed information on federal inventory reporting requirements and calculation methodologies. Both documents were circulated by the U.S. Council on Environmental Quality for public comment in July 2010.

Skepticism and scientific scandal

In 2009, hackers released a large cache of e-mails from East Anglia University's Climate Research Unit (CRU), which fueled the greenhouse controversy and allegations of scientific fraud. Some of these e-mails are indicative of a long and concerted history of scientific misconduct. It began with Geoff Jenkins, chairman of the IPCC's first scientific group, who admitted in 1996 to a "cunning plan" to feed fake temperature data provided to Nick Nuttall, head of media for the UN's environmental program.

Phil Jones, the CRU's director, e-mailed instructions to "hide the decline" in a recently observed cooling trend. Jones boasted in one e-mail that he had used a statistical "trick" to do just that.

Michael Mann was the director of Penn State University's Earth System Science Center, a principal author of the IPCC, and developed the "hockey stick" graph purporting to prove that global temperatures were at the hottest in recorded history. But in 2008, statistician Steve McIntrey exposed the flawed methodology. Other climate scientists were incredulous that Mann's work had passed two IPCC peer-reviewed rounds without anyone spotting these inaccuracies. Andrew Revkin, a senior *New York Times* reporter inquired about Mann's discredited analysis; Mann replied in an e-mail to Revkin that "Those...who operate outside ... the system are not to be trusted." Instead of investigating the truth, Revkin supported Mann's rebuttal. When the CRU's incriminating e-mails came to light, Revkin retorted that they had been "stolen." Revkin's reply was indeed interesting, given that the *New York Times* has a long history of printing leaked national security secrets.

Other e-mails reveal efforts to ostracize skeptical researchers from the scientific community. Research that was inconveniently disagreeable to the CRU's mindset was ignored. Efforts were taken to delete opposing research from IPCC reports. Some administrators had threatened to boycott journals that dared print research papers showing evidence to the contrary. Efforts were even taken to withhold data and even delete some data that researchers did not want revealed to the public.

Perhaps most damaging of all is the fact that the CRU maintained an important database of information, critical to the IPCC analyses and conclusions. The original data measurements used in the IPCC reports had been "corrected." This, in and of itself, was not necessarily improper, as data correction techniques are commonly employed in scientific analyses to compensate for errors. However, the original uncorrected dataset was destroyed, purportedly because there was insufficient room to store it. At the very least, this is scientific sloppiness at its worst. But it implies something of profound importance to the science of climate change: There is now no means of verifying the accuracy of much of the original uncorrected data. This is particularly troubling as it calls into question data that are critical to the case of global warming, and upon which world economies are gambling trillions of dollars over the coming decades.

Critics have dubbed this scientific coup "climategate." Sadly, such scientific misconduct is the modern equivalent of scientific heresy. It has cast a shadow of suspicion over the discipline of climate change research. How far this scandal spreads, as of now, is anyone's guess. It may take years for ethical scientists to reestablish the trust that has been lost among the public and policymakers.

Before presenting guidance on addressing GHG impacts in NEPA documents, we will briefly introduce a summary of key NEPA court cases involving climate change.

Summary of key NEPA court decisions involving climate change

This section summarizes some recent US lawsuits involving climate change issues in NEPA documents. The reader is cautioned that case law is evolving and should consult with legal counsel in applying these cases to a specific GHG assessment.

An early important case

An early legal challenge involving GHG emissions is the 1990 case of *City of Los Angeles v. National Highway Traffic Safety Administration*.[9] This case concerned the setting of the US Corporate Average Fuel Economy (CAFE) standard. The suit alleged that a lower standard would adversely affect global warming. It is notable for two reasons:

1. The court held that the plaintiffs (parties filing the lawsuit) had legal *standing* to bring the lawsuit (a significant holding in its own right).
2. A 1-mile per gallon change in the CAFE standard at issue was not so significant as to require an EIS.

What is of importance is that this court (like nearly all subsequent federal courts to address the climate change issue) did not doubt that global warming was a proper subject for analysis under the NEPA; the court merely concluded that the impacts of this particular action fell below the threshold of significance requiring preparation of an EIS.

In some cases, a project has been challenged even if it did not directly produce GHG emissions. For example, one court has already ruled that a proposal restricted to construction of electrical transmission and railroad lines for transporting coal to the power plants required an analysis of the indirect impacts on CO_2 that would be generated by these plants.[10]

In addition to GHG emissions, NEPA analyses also may (and probably should) be required to consider the consequences of climatic change (e.g., from rising sea levels, increased storm activity, severe weather, flooding, or reduced access to water). If direct and indirect impacts of a pollutant as ubiquitous as CO_2 must be analyzed, along with reasonably foreseeable impacts of climatic change, the depth and scope of the analysis for many proposals could be substantially expanded beyond that which would have been deemed previously acceptable. Although courts have yet to reject an

EIS (providing some degree of GHG analysis was included) on grounds that its analysis of climate change was inadequate, it is a risk that federal agencies must consider as events continue to unfold.

To date, most challenges based on climate change have centered primarily on the relative lack of any GHG emission computations or analysis in an environmental assessment (EA) or EIS. The results of these cases have been mixed. In one of the most notable cases, *Mayo Foundation v. Surface Transportation Board*, the court concluded it was unnecessary to impose additional mitigating conditions on minor increases (less than 1%) in CO_2 emissions associated with increased combustion of coal that would result from a railroad expansion proposal.[11] In contrast, the US Ninth Circuit Court recently concluded in *Center for Biological Diversity v. Nat'l. Highway Traffic Safety Administration* that the "impact of greenhouse gas emissions on climate change is precisely the kind of cumulative impacts analysis that NEPA requires agencies to conduct."[12]

While a number of federal court cases have required agencies to consider GHG emissions in their NEPA documents, with respect to their adequacy, the courts have tended to defer in favor of an agency's internal expertise and assessment. However, some lawyers believe that courts will become increasingly more demanding in their expectations. To date, the Ninth Circuit has been the most demanding court in terms of its assessment expectations.

Ninth Circuit guidance

In reaching its decision concerning the need to address GHG emissions in NEPA documents, the US Ninth Circuit Court summarized findings from sources such as the IPCC report (Table 2.4).

Based on scientific evidence that the Ninth Circuit cited in its ruling, this court clearly considers climate change an appropriate issue in NEPA documents.

Litigation strategies used by plaintiffs

Plaintiffs have used various strategies on which to base their climate change cases, including arguments that

- Alternatives to the proposal with a lower potential to emit GHGs were not adequately evaluated.
- GHG emissions were reasonably foreseeable significant indirect and cumulative impacts requiring preparation of an EIS.
- GHG impacts involving incomplete or unavailable information should have been prepared consistent with NEPA's provision for addressing such uncertainty (40 CFR §1502.22). Note: see Table 2.11.

Table 2.4 Ninth Circuit Findings from Sources such as the International Panel
on Climate Change Report

The average earth surface temperature has increased.

CO_2 concentrations increasing over the twenty-first century are virtually certain
to be primarily the result of fossil-fuel emissions.

Global warming will affect plants, animals, and ecosystems around the world.
Some scientists predict that it will cause 15% to 37% of species in certain regions
to become extinct.

There have been severe impacts on the Arctic due to warming, including the
melting of sea ice.

Global warming will cause serious consequences for human health, including the
spread of infections and respiratory diseases.

Climate change may be nonlinear, meaning there are positive feedback
mechanisms that may push global warming past a dangerous threshold (the
"tipping point").

Climate change is associated with increasing variability and heightened intensity
of storms and hurricanes.

Plaintiffs have challenged projects for their potential to (1) change the
Earth's global climate, as well as how (2) climatic change could affect specific environmental resources. For instance, in *Border Power Plant Working Group v. Department of Energy* (discussed in more detail shortly), the US Department of Energy (DOE) was challenged for failing to analyze potentially significant changes that a project might have upon the Earth's climatic system.[13] Similarly, in *Center for Biological Diversity v. Kempthorne*[14] (discussed in more detail shortly), plaintiffs alleged that the US Fish and Wildlife Service failed to analyze the effects of global warming on polar bears and walruses when it adopted a final rule authorizing the incidental taking of these species.[15]

The next section reviews some recent court cases, which shed some important light on the extent to which GHG emission and climate change analyses should be included in NEPA documents.

Lessons learned from climate change litigation

To date, courts have not been shy about addressing GHG emissions and global climate change issues in NEPA documents. Such projects have included federal permitting decisions,[16] federal rulemaking[17] (including CAFE standard setting[18]), federally approved construction projects,[19] federal leases,[20] and the federal financing of projects.[21]

Common law tort claims such as nuisance and negligence are also gaining greater attention from litigants claiming global warming injuries. For example, in *California v. General Motors Corporation*,[22] the Attorney General of California filed suit against six major auto manufacturers for damages, including "future monetary expenses and damages as may be incurred by California in connection with the nuisance of global warming." With mounting scientific evidence in support of global warming and its causes, tort claims based on negligence may become increasingly common. Described below are some recent US court cases involving GHG emission impacts. The reader is cautioned to consult with legal counsel in determining the applicability of these cases to specific projects.

Cases holding EA to be inadequate
This section describes an important case in which an EA was found to be invalid, based on climate change issues.

Center for Biological Diversity v. NHTSA 508 F3d. 508 (9th Cir. 2007) (rehearing petition pending)   In this case, the court held that the National Highway Traffic Safety Administration (NHTSA) needed to prepare a full EIS addressing the effects that its fuel economy standards would have on global climate. Plaintiffs alleged, in part, that the NHTSA's EA was inadequate because it failed to take *a hard look* at the GHG implications of its rulemaking. Additionally, the suit alleged that the agency failed to analyze a reasonable range of alternatives and did not adequately examine the cumulative impact of the proposed rule.

The court ruled that where a substantial question exists as to whether an action may have a significant environmental effect, the agency must prepare an EIS. In the court's words, "there is no doubt that the fuel economy standards set by NHTSA will have a direct effect on greenhouse gas emissions from light trucks, and that NHTSA is a 'legally relevant cause.'"

The NHTSA concluded that the CO_2 emitted as a result of its gas mileage rule would not have a significant cumulative impact on the environment. Its conclusion was based on its EA, which calculated the total tonnage of such emissions under the CAFE standards; the calculation showed a reduction in CO_2 emissions of between 122 to 196 million metric tons, compared to the tonnage that would result if the rule were not implemented. Thus, the NHTSA concluded that the new standards were an improvement over the status quo and that no significant impact would occur.

The Ninth Circuit Court emphatically declared that the effect of GHG emissions "on climate change is precisely the kind of cumulative impact analysis that NEPA requires agencies to conduct." While the new gas mileage rules might constitute an improvement over the status quo, adopting them would only slow the rate of growth of CO_2 emissions. The court's holding on GHG impacts was unanimous. The court concluded the EA analysis and the finding of no significant impact (FONSI) based on it were inadequate, particularly in their failure to consider the cumulative effect of the new fuel standards on climate change. While the court acknowledged that the proposed action was "an improvement" over previous CAFE standards, it also noted that this standard would nevertheless result in a "significant effect." The NHTSA subsequently issued a final EIS in 2010.

This case is important because the court made four observations outlined in Table 2.5. This case is particularly noteworthy because of the weight the court gave to including a rigorous cumulative CO_2 impact analysis in an NEPA document.

Cases in which an EA suit is currently pending

Described below are important cases in which EAs involving climate change issues are currently pending.

Ranchers Cattleman Action Legal Fund v. Conner[23]   In this case, plaintiffs challenged the United States Department of Agriculture's (USDA) promulgation of regulations relaxing restrictions on importing live cattle and edible bovine products from minimal risk mad cow disease regions (i.e., Canada). The suit claims that the EA was inadequate because it did not analyze the increased emissions of GHGs associated with transportation and importation of cattle into the United States.

Table 2.5 Four Principal Points Made by the Ninth Circuit Court in *Center for Biological Diversity v. NHTSA*

- The NEPA requires federal agencies to prepare an EIS if there is a substantial question as to whether a proposal may result in a significant effect, either individually or cumulatively. After reviewing the NHTSA's EA, the court concluded that the cumulative impacts analysis was inadequate because the project might produce emissions that would result in a significant environmental effect.

- The court noted that, while the EA quantified the anticipated CO_2 emissions from light trucks that would be governed under the proposed rule, it did not evaluate the incremental impact of these emissions on climate change, particularly in terms of cumulative impacts (e.g., other past, present, and reasonably foreseeable actions).

- The court went on to declare that while climate change is a global phenomenon that includes actions that lie beyond an agency's control, this fact does not release the agency from the duty of assessing the effects of its action on global warming.

- Finally, the court rejected the agency's argument that an EIS was not required because the new standards would result in decreased CO_2 emissions related to the older standards. The court concluded that simply because the final rule may be viewed as an improvement over the existing standards does not imply that it would have no significant impact. In reaching this conclusion, the court noted that simply accounting for GHGs generated by the proposed rule does not constitute a "hard look" at the environmental consequences. Thus, the court directed NHTSA to prepare an EIS to analyze the impact of GHG emissions on climate change.

Center for Biological Diversity v. Kempthorne[24]   As described earlier, plaintiffs challenged an EA for a US Fish and Wildlife Service regulation allowing for the incidental taking of polar bears and Pacific walruses associated with new oil and gas exploration and development in the Arctic. The suit alleges that the EA is inadequate because it fails to evaluate and quantify the cumulative impact of anticipated industrial activities and how they will exacerbate harm to polar bears and walruses already threatened by climate change.

Case holding an EIS to be invalid

Described below is an important case in which an EIS involving climate change issues was found to be invalid.

Mid-states Coalition for Progress v. Surface Transportation Board[25]   The Surface Transportation Board approved construction of approximately 280 miles of new rail line to reach Wyoming coal mines and to upgrade the existing rail line. The board prepared an EIS to examine the effects of constructing and operating the rail line to accommodate coal traffic expected to result from the project.

The Sierra Club challenged the action, arguing that the board failed to consider the effects on air quality that an increased supply of low-sulfur coal for future power plants would produce. The defendants argued that if the availability of coal would drive the construction of additional power plants, the board would need to know where those plants would be built, and how much coal these new plants would consume. Because no hauling contracts had been executed and the relevant information was unknown, the board argued that such an analysis would amount to "pure speculation."

The court held that in such circumstances, federal agencies are directed to follow the CEQ procedure for evaluating reasonably foreseeable significant adverse effects when there is "incomplete or unavailable information" (40 CFR §1502.22).

In explaining why it did not include an analysis of air emissions, the defendants argued that the 1990 *Clean Air Act* Amendments mandate reduction in pollutant emissions; an assumption in the analysis was that emissions would definitely fall to the mandated level, reducing whatever effect the emissions will have on global warming. The court recognized that this assumption might be true for pollutants capped under the *Clean Air Act*, but does not apply to pollutants that have not been capped.

On remand, the agency prepared a supplemental EIS complying with the NEPA's regulatory provisions for dealing with incomplete or unavailable information, which was subsequently upheld by the US Eighth Circuit Court.[26]

Cases holding EISs to be valid

Described below is a summary of important cases in which EISs involving climate change issues were found to be valid.

Seattle Audubon Society v. Lyons[27]   Environmental groups and the timber industry challenged the legality of a forest management plan. One of the plaintiff's claims was that the supplemental EIS failed to disclose the impacts of timber harvest on water quality, air quality, and *climate*. The court held that the final supplemental EIS had discussed these impacts at length and therefore was valid in that respect.

Association of Public Agency Customers v. Bonneville Power Administration[28]   The Bonneville Power Administration adopted a market-driven business plan, executing sale contracts with direct-service industrial customers. This plan would increase the demand for more power. In this case, the NEPA analysis briefly addressed GHG emissions pursuant to Executive Order 12114, requiring federal agencies to develop procedures that take extraterritorial impacts on the global commons into account for major federal proposals. Plaintiffs argued that the Bonneville Power Administration's EIS did not adequately discuss global warming implications from the effects of GHGs released as a result of increased operations, and did not evaluate transboundary impacts on Canada. The court ruled that the EIS sufficiently considered these issues.

Mayo Foundation v. Surface Transportation Board[29]   As described earlier, this case involved approval of new railroad lines for transporting low-sulfur coal from Wyoming to Midwest power plants. The Eighth Circuit Court initially ruled that increased coal consumption and related GHG emissions were a reasonably foreseeable consequence and that the Surface Transportation Board should have considered such issues in the EIS.

However, the court upheld a supplemental EIS in December 2006, which concluded that the project would not result in significant GHG impacts. The EIS estimated that the project would increase global GHG emissions by 0.088%, and United States' GHG emissions by 0.023%; the EIS concluded that the estimated impact upon global climate change would be negligible. The board also concluded that it was not necessary to impose additional mitigating conditions on the project. The Eighth Circuit Court rejected the Sierra Club's argument that the analysis was inadequate, noting that the board more than adequately considered the "reasonably foreseeable" adverse effects.

EIS suit currently pending

This section summarizes a case in which an EIS involving climate change issues is currently pending.

Montana Environmental Information Center v. Johanns[30]   The USDA's Rural Utility Service's (RUS) use of low-interest loans to help finance construction of at least eight new coal-fired power plants was challenged. The plaintiffs assert that the RUS-funded projects will account for a "significant share" of United States GHG emissions, yet the EIS failed to take a "hard look" at the consequences of such a major federal action. Specifically, they alleged that RUS

- Failed to consider the cumulative or incremental impacts of GHG emissions from other coal plants that it was considering for funding
- Failed to consider a reasonable range of alternatives
- Should have prepared a supplemental EIS based upon new information that was received after the issuance of the EIS

The case is currently pending.

Non-NEPA suits supporting consideration of global climate change

Described below are some additional non-NEPA lawsuits that have important implications on global climate change issues.

Clean Air Act   As described earlier, the Supreme Court ruled in *Massachusetts v. EPA,* that the plaintiff had standing to bring a suit, which claimed that a rise in sea level associated with global warming had already harmed and will continue to harm the state of Massachusetts. The Supreme Court found that the Clean Air Act authorizes the EPA to regulate GHGs from new motor vehicles. The court remanded the case to the EPA to determine whether GHGs might endanger public health or welfare, and therefore should be regulated. The EPA has begun the formal process for this review.

Endangered Species Act   As described earlier, *Natural Resources Defense Council v. Kempthorne*[31] concerns a large water diversion project in California. The US Fish and Wildlife Service assumed that the hydrology of the water bodies affected by the project would follow historical patterns over the next two decades. However, it appears that potential changes in climate resulting from GHG emissions might produce earlier flows, more floods, and drier summers. The court found that it was arbitrary and capricious for the Fish and Wildlife Service to ignore this evidence.

Clean Water Act   In 2007, an advocacy group filed a petition involving eight states, requesting that they declare their coastal waters "impaired" by CO_2 emissions under the Clean Water Act. The aim was to force states to develop a water pollution standard for CO_2 (which turns water acidic) under the Clean Water Act's non-point source provisions, and to limit emissions to achieve that standard.

Global Change Research Act   This U.S. 1990 act requires a federal scientific body to prepare periodic scientific assessments on global climate change. A suit was brought by several environmental groups, led by the Center for Biological Diversity (CBD). In 2007, a district court found that the federal defendants had not filed the required reports and ordered them to do so.[32]

Regulatory direction on considering greenhouse emissions

This section describes some significant regulatory direction that is useful in determining if and how greenhouse emissions should be examined in NEPA documents.

Focus of current guidance

Five distinct types of activities depicted in Table 2.6 have been described in guidance and scientific literature for assessing climate change impacts in NEPA analyses. Potential impacts resulting from these categories of activities may affect the "human environment" in general. These impacts include disruptions resulting from temperature variations, drought and changes in snowpack, rising sea levels, changing water tables, and increased flooding. While the emissions are relatively easy to calculate given the state of the art, assessing the resulting climate impacts can be a problematic undertaking.

Executive Order 13423

On January 24, 2007, President George W. Bush signed Executive Order (E.O.) 13423.[33] This order set goals in the areas of energy efficiency, acquisition, renewable energy, toxic pollutant reductions, renewable energy, sustainable buildings, fleets, and water conservation. In addition, it mandated greater use of environmental management systems (EMSs) as a mechanism for managing and continually improving sustainable practices.

Under this order, through life-cycle cost-effective energy measures, each agency was instructed to reduce its GHG emissions attributed to facility energy use by 30% by 2010 compared to 1990 emission levels. To encourage optimal investment in energy improvements, agencies could count GHG reductions from improvements in non-facility energy use toward this goal to the extent that these reductions are approved by the Office of Management and Budget (OMB).

Table 2.6 Guidance for Assessing Climate Change Impacts in NEPA Analyses Focusing on Five Broad Categories of Activities

Power plant emissions include GHGs emissions produced principally by fossil-fueled power plants. This category may also include offsets produced by non-fossil-fueled power plants.

Purchased electricity emissions include GHGs emissions produced when power is purchased from a plant, which generated the electricity at another site. Computer models can be used to estimate emissions based on energy usage or from various types of facilities. The projected purchase of electrical power is multiplied by a prescribed emission factor, which estimates the CO_2 emissions.

General construction emissions include accounting for GHG construction emissions. Types of activities include construction equipment use and fabrication of construction materials such as cement, which can produce large volumes of CO_2 off-gases.

General operational emissions: These impacts typically include:

Facility stack emissions

Fugitive emissions from oil and gas wells

Methane from landfills

Methane from wastewater treatment plants

Impacts such as uptake of solid carbon or CO_2 emissions associated with carbon sinks (wetlands, forests, and agricultural operations)

Emissions of methane and nitrous oxide from agricultural operations

Transportation emissions: The impacts include employee travel, and transportation of materials and manufactured goods and indirect actions such as transporting coal by rail to fuel power plants in another region. Modeling software can be used to compute projected vehicle travel miles, which can then be multiplied by specific emission factors.

Federal leadership in environmental, energy, and economic performance

E.O. 13514, *Federal Leadership in Environmental, Energy, and Economic Performance,* directs U.S. agencies to establish an integrated strategy for sustainability and make reduction of GHG emissions a federal agency priority.[34] Under this E.O., agencies are required to set specific targets for reducing GHG emissions and adopt measures to attain those targets. The E.O. further directs agencies to enhance other aspects of sustainability by reducing water consumption, minimizing waste, supporting sustainable communities, and using federal purchasing power to promote environmentally responsible products and technologies.

E.O. 13514 requires agencies to establish reduction targets for various types of GHG sources. Agencies must track and report their progress annually to the chair of the CEQ and the director of the OMB on three GHG categories:

1. Direct emissions from sources owned or controlled by the agency
2. Direct emissions from generation of electricity, heat, or steam purchased by the agency
3. Emissions from sources not owned or controlled by the agency but related to agency activities, such as vendor supply chains, delivery services, and employee travel and commuting

E.O. 13514 requires the analysis of energy consumption in all EAs and EISs for proposals involving new or expanded federal facilities. This E.O. directs agencies to ensure that planning for new federal facilities or new leases includes consideration of sites that are pedestrian-friendly, near existing employment centers, and accessible to public transit.

Congress requires EPA to create GHG emissions reporting regulation

On December 26, 2007, President Bush signed into law an omnibus spending bill, H.R. 2764, for 2008. This bill instructed the US EPA to create a GHG emissions registry and a GHG reporting regulation for emissions that exceed certain thresholds. The final rule was published in December 2009. This reporting requirement applies to economy-wide activities. EPA is required to determine how often industries must submit reports, and it must also consider reporting requirements for both upstream and downstream sources of production.

NEPA and GHG impact considerations

As scientists generate more evidence about the phenomenon of climate change, agencies are beginning to place emphasis on determining appropriate methods for assessing GHG emission impacts in NEPA. Increasingly, other nations are also including GHG assessments in their EIAs. Key GHG assessment issues consider whether:

- The proposed action contributes directly or indirectly to GHG emissions.
- The proposal could result in a cumulative GHG emission (consideration of the impacts of the proposed action, when combined with other past, present, and reasonably foreseeable actions).
- Sufficient information is available to describe the nature and extent of the proposed action's effect.

NEPA analyses offer perhaps the single most efficient mechanism for integrating GHG considerations into the federal decision-making process.

Specifically, NEPA offers an integrated federal planning process for evaluating GHG emissions and their potential impacts, and analyzing alternatives and mitigation measures for implementing courses of action with lower carbon footprints. Many other nations have adopted similar EIA processes that can also be used to combat the impacts of GHG emissions.

State of current NEPA practice

As witnessed earlier, most NEPA documents, to date, have included little or no consideration of potential climate change impacts. Agencies have offered many excuses for neglecting such issues. Table 2.7 presents some of the common excuses for not addressing GHG emission impacts in NEPA documents.

The author anticipates that in the near future, failure to consider potential GHG emissions will result in many more legal challenges.

Increased regulatory and EPA oversight

One of the clearest indications that courts are willing to impose GHG emission assessments at the local level involves a recent US Supreme Court case ruling in *Massachusetts v. EPA*.[35] Challenged for refusing to regulate CO_2 emissions, the EPA argued that it lacked both the authority and obligation to regulate GHGs. The Supreme Court accepted the scientific findings of the IPCC, and held that "[t]he harms associated with climate change are serious and well recognized." The court ruled that GHGs such as CO_2 are pollutants subject to regulation under the Clean Air Act. Based on this ruling, NEPA decision makers should seriously consider the extent to

Table 2.7 Common Excuses for Not Addressing GHG Emission Impacts in NEPA Documents

We've never addressed these impacts before, so there is no precedent for doing so now.

Typical computational methods are not sophisticated enough to evaluate such impacts.

Climate change is a global phenomenon, and NEPA does not have to address such extraterritorial issues.

Greenhouse gases are not regulated by the EPA.

Such impacts are beyond the scope of this analysis.

There is no scientific consensus on greenhouse warming, making it impossible to perform an analysis or reach conclusions.

The issue is beyond the state-of-art, therefore this amounts to attempting to address remote and speculative impacts which do not have to be evaluated under NEPA.

The problem is insurmountable, so why try?

which they may be responsible for identifying, evaluating, and disclosing potential GHG emission impacts, including mitigation measures.

Consistent with the Supreme Court ruling, the EPA has begun to comment on the GHG issues as part of its EIS review responsibility. For example, in 2007, EPA Region VIII criticized a draft EIS involving mining activities conducted by the Mountain Coal Company at its West Elk Mine in Colorado. In this case, the Forest Service prepared an EIS for a proposal to drill drainage wells and a ventilation shaft to release methane. The EPA criticized the draft EIS for failing to predict and assess the methane emission—a GHG more than 20 times as potent as CO_2.[36] The EPA also concluded that the EIS did not adequately analyze capture and utilization of methane as an energy resource. The EPA scored the draft document as "insufficient," emphasizing that the missing analyses were substantial issues that needed to be addressed in the final EIS.

Similarly, the US Department of Interior has signaled increased interest in the assessment of climate change impacts in NEPA documents. The Department of Interior recently created an internal task force on climate change to study climate science, and land and water management, with emphasis on assessing the implications of climate change for a range of documents the agency relies on in reaching decisions (e.g., resource management plans and NEPA documents).

The NEPA's public element

It should be emphasized that litigation is not the only course available to those wishing to have GHG emissions addressed in NEPA documents. NEPA offers numerous opportunities for public participation. The EIS (and sometimes the EA) process can be noisy by offering numerous political and regulatory mandated pressure points for compelling an agency to address GHG emissions. For instance, the scoping process, in which interested persons can offer suggestions on the contents of the EIS (EAs are sometimes also subject to public review of the draft document), is one avenue in which to focus attention on GHG issues. Another opportunity is the draft EIS comment period.

Draft CEQ guidance on considering climate change and greenhouse gas

A draft memorandum for public comment was issued by the CEQ in 2010. It provides guidance for assisting federal agencies in improving their consideration of the effects of GHG emissions and climate change impacts in their NEPA analyses.[37] This draft guidance is not intended as a new component of NEPA analysis, but rather as a potentially important factor to be considered within the existing NEPA framework. This draft guidance is summarized below.

Because climate change is a global problem that results from global GHG emissions, there are more sources and actions emitting GHGs (in terms of both absolute numbers and types) than are typically encountered when evaluating the emissions of other pollutants. From a quantitative perspective, there are no dominating sources and fewer sources that would even be close to dominating total GHG emissions. The global climate change problem is much more the result of varied sources, each of which might seem to make a relatively small addition to global atmospheric GHG concentrations. The CEQ proposes to recommend that environmental documents reflect this global context and be realistic in focusing on ensuring that useful information is provided to decision makers for those actions that the agency finds are a significant source of GHGs.

Under this proposed guidance, agencies should use the scoping process to set reasonable spatial and temporal boundaries for this assessment and focus on aspects of climate change that may lead to changes in the impacts, sustainability, vulnerability, and design of the proposed action and alternative courses of action. At the same time, agencies should recognize the scientific limits of their ability to accurately predict climate change effects. Where an agency determines that an assessment of climate issues is appropriate, the agency should identify and assess alternative actions that are both adapted to anticipated climate change impacts and mitigate the GHG emissions that cause climate change.

When to evaluate GHG emissions

Where a proposed federal action that is analyzed in an EA or EIS would be anticipated to emit GHGs to the atmosphere in quantities that the agency finds meaningful, it is appropriate for the agency to quantify and disclose its estimate of the expected annual direct and indirect GHG emissions in the environmental documentation for the proposed action.

De Minimis assessments   The term *de minimis* refers to a sufficiently small threshold concentration or level that has been demonstrated to be of no substantive concern, and is therefore exempt from further regulatory analysis. For the purposes of its draft guidance, the CEQ defines GHGs in accordance with Section 19(i) of Executive Order 13514. The CEQ proposes to advise federal agencies to consider, in scoping their NEPA analyses, whether analysis of the direct and indirect GHG emissions from their proposed actions may provide meaningful information to decision makers and the public. Specifically, if a *stationary source* would directly emit 25,000 metric tons or more of CO_2-equivalent GHG emissions on an *annual basis*, agencies should consider this an indicator that a quantitative or qualitative NEPA analysis may be meaningful to decision makers and the public.

For long-term actions that have annual direct emissions of fewer than 25,000 metric tons of CO_2-equivalent, the CEQ encourages federal agencies to consider whether the action's long-term emissions should receive similar analysis.

The figure of 25,000 metric tons may provide a useful, presumptive threshold for discussion and disclosure of GHG emissions because it has been used and proposed in rulemakings under the Clean Air Act[38] because it provides comprehensive coverage of emissions with a reasonable number of reporters, thereby creating an important data set useful in quantitative analyses of GHG policies, programs, and regulations.[39] This rationale is pertinent to the presentation of NEPA analyses as well.

The CEQ does not propose this reference point as an indicator of a level of GHG emissions that may significantly affect the quality of the human environment, as that term is used by NEPA, but notes that it serves as a minimum standard for reporting emissions under the Clean Air Act. Evaluation of significance under NEPA is done by the action agency based on the categorization of actions in agency NEPA procedures and action-specific analysis of the context and intensity of the environmental impact.[40] Examples of proposals for federal agency action that may warrant a discussion of the GHG impacts of various alternatives, as well as possible measures to mitigate climate change impacts, include approval of a large solid waste landfill, approval of energy facilities such as a coal-fired power plant, or authorization of a methane venting coal mine.

Quantifying emissions   Once an agency has determined that this step is appropriate, the CEQ proposes that agencies should consider quantifying emissions using the following technical documents, to the extent that this information is useful and appropriate for the proposed action under NEPA:

- Quantification of emissions from large direct emitters: 40 CFR Parts 86, 87, 89, et al. Mandatory Reporting of Greenhouse Gases; Final Rule, US Environmental Protection Agency (74 Fed. Reg. 56259-56308). Note that "applicability tools" are available (http://www.epa.gov/climatechange/emissions/GHG-calculator/) for determining whether projects or actions exceed the 25,000 metric tons of CO_2-equivalent greenhouse gas emissions.
- Quantification of Scope 1 emissions at federal facilities: GHG emissions accounting and reporting guidance that will be issued under Executive Order 13514 Sections 5(a) and 9(b) (http://www.ofee.gov).
- Quantification of emissions and removals from terrestrial carbon sequestration and various other project types: Technical Guidelines, Voluntary Reporting of Greenhouse Gases, (1605(b) Program, US Department of Energy [http://www.eia.doe.gov/oiaf/1605/]).

Land management   Land management techniques, including changes in land use or land management strategies, lack any established federal protocol for assessing their effect on atmospheric carbon release and sequestration at a landscape scale. Therefore, CEQ is seeking public input on this issue.

What should be considered in the GHG evaluation

The following guidance is offered for considering GHG assessments in NEPA documents.

Rule of reason   The NEPA is governed by the rule of reason which ensures that agencies determine whether and to what extent to prepare an EIS based on the usefulness of any new potential information to the decision-making process.[41] Agencies apply the rule to ensure that their discussion pertains to the issues that deserve study and de-emphasizes issues that are less useful to the decision regarding the proposal, its alternatives, and mitigation options.[42] The draft memorandum recommends that agencies ensure that their emission and impact descriptions are commensurate with the importance of the GHG emissions of the proposed action, avoiding useless bulk and boilerplate documentation, so that the NEPA can concentrate on important issues.[43]

Where a proposed action is evaluated in either an EA or an EIS, the agency may look to reporting thresholds in the technical documents cited above as a point of reference for determining the extent of direct GHG emissions analysis that is appropriate to the proposed agency decision. As proposed in the draft guidance above, for federal actions that require an EA or EIS, the direct and indirect GHG emissions from the action should be considered in scoping and, to the extent that scoping indicates that GHG emissions warrant consideration by the decision maker, quantified and disclosed in the environmental document.[44] In assessing direct emissions, an agency should look at the consequences of actions over which it has control or authority.[45]

Cumulative effects   Where an agency concludes that a discussion of cumulative effects of GHG emissions related to a proposed action is warranted to informed decision making, the CEQ recommends that the NEPA analysis focus on evaluating the annual and cumulative emissions of the proposed action and the difference in emissions associated with alternative actions.

Programmatic analysis   An agency may find it useful to describe GHG emissions in aggregate, as part of a programmatic analysis EA or EIS that evaluates activities which can be incorporated by reference into subsequent NEPA analyses for individual project-specific agency actions.

Federal programs that affect emissions or sinks, and proposals regarding long-range energy, transportation, and resource management programs lend themselves to a programmatic approach.

Mitigation To the extent that a federal agency evaluates proposed mitigation of GHG emissions, the quality of that mitigation—including its permanence, verifiability, enforceability, and *additionality*—should also be carefully evaluated. Regulatory additionality requirements are designed to ensure that GHG reduction credit is limited to an entity with emission reductions that are above regulatory requirements.[46] Among the alternatives that may be considered for their ability to reduce or mitigate GHG emissions are enhanced energy efficiency, lower GHG-emitting technology, renewable energy, planning for carbon capture and sequestration, and capturing or beneficially using fugitive methane emissions.

Assessing climate change impacts Agencies should determine which climate change impacts must warrant consideration in their EAs and EISs. Through scoping, agencies determine whether climate change considerations warrant emphasis or de-emphasis.[47] When scoping the impact of climate change on the proposal for agency action, the sensitivity, location, and time frame of a proposed action will determine the degree to which consideration of these predictions or projections is warranted. As with the analysis of any other present or future environment or resource condition, the observed and projected effects of climate change that warrant consideration are most appropriately described as part of the current and future state of the proposed action's "affected environment."[48]

Level of detail The level of detail in any environmental analysis should be commensurate with the rule of reason. Agencies should ensure that they keep in proportion the extent to which they document their assessment of the effects of climate change. The focus of this analysis should be on the aspects of the environment that are most affected by the proposed action.

The level of detail in the analysis of these effects will vary among affected resource values. For example, if a proposed project requires the use of significant quantities of water, changes in water availability associated with climate change may need to be discussed in greater detail than other consequences of climate change.

Identifying reasonably foreseeable future conditions When assessing the effects of climate change on a proposed action, an agency typically starts with an identification of the reasonably foreseeable future condition of the affected environment for the no-action alternative based on available climate change measurements, statistics, observations, and other evidence.

The reasonably foreseeable affected environment should serve as the basis for evaluating and comparing the incremental effects of alternatives.[49] Agencies should be clear about the basis for projecting the changes from the existing environment to the reasonably foreseeable affected environment, including what would happen under this scenario and the probability or likelihood of this future condition.

The obligation of an agency to discuss particular effects turns on "a reasonably close causal relationship between the environmental effect and the alleged cause."[50] Where climate change effects are likely to be important but there is significant uncertainty about such effects, it may also be useful to consider the effects of any proposed action or its alternatives against a baseline of reasonably foreseeable future conditions that is drawn as distinctly as the science of climate change effects will support.

Monitoring In cases where adaptation to the effects of climate change is important, the significant aspects of these changes should be identified in the agency's final decision, and adoption of a monitoring program should be considered. Monitoring strategies should be modified as more information becomes available and best practices and other experiences are shared.

In accordance with NEPA's rule of reason and standards for obtaining information regarding reasonably foreseeable significant adverse effects on the human environment, action agencies need not undertake elaborate research or analysis of projected climate change impacts in the project area or on the project itself, but may instead summarize and incorporate by reference the relevant scientific literature.[51]

Criteria useful in determining the need to evaluate

GHG impacts in NEPA documents

As outlined below, Schussman et al. have identified some criteria useful in determining the need to evaluate climate change effects in NEPA documents.[52] These criteria are described below.

Uncertainty

Given the degree of scientific uncertainty surrounding GHG emissions and their effect on climate change, any NEPA analysis will likewise involve uncertainties, including gaps in information. Such deficiencies, however, do not necessarily exempt an agency from responsibility to perform such an analysis. This criterion is reinforced by the following case.

Mid-states Coalition for Progress v. Surface Transportation Board[53]   As described earlier, the US Eighth Circuit Court found that the agency acknowledged that increased pollutant emissions (including CO_2) would occur. However, the NEPA analysis did not evaluate such emissions because information was not available as to the specific location or attributes of the emitting facilities.

The court reasoned that the nature of the effect could be known, but not the extent of the impact; it concluded that NEPA requires an analysis to be prepared in accordance with 40 CFR §1502.22 (incomplete or unavailable information). While this case may impact a small number of coal-fired power plants, the outcomes could have far-reaching consequences on an array of proposals, including but not limited to private projects with GHG emissions that require federal approvals.

The causal chain and reasonably close relationship

In deciding whether to analyze a specific impact under NEPA (including climate change), agencies first need to consider the extent to which such effects would be *caused* by the proposed federal action. In general, only those impacts that bear a "reasonably close"[54] relationship to the major federal action that is the subject of the EIS fall within the reach of NEPA—a relationship akin to the *tort doctrine of proximate cause*. The length of NEPA's causal chain[55] is determined by examining the "underlying policies or legislative intent"[56] and considering the "legal responsibility of actors."[57]

In at least one case, the court did not decide whether the action was a legally relevant (see next section) cause of the alleged climate change. This was based on the fact that the record did not contain sufficient evidence that the defendant's action was a "but-for" cause of project emissions; moreover, the record did not demonstrate that the defendant had sufficient regulatory control over the project emissions. The causality criterion is reinforced by the following cases.

Center for Biological Diversity v. NHTSA[58]   As described earlier, this case considered whether combustion of fossil fuels from a proposed project could substantially contribute to climate change. A key question was whether, under NEPA, the federal action was a "legally relevant cause" of the effect. Such a determination may depend on

- The nature of the proposed action
- Whether the project at issue (or the relevant greenhouse gas emissions) would occur regardless of the federal action
- The amount of federal agency control over the project or GHG emissions
- The extent of the anticipated effect

The US Ninth Circuit Court found that a federal proposal to set corporate average fuel economy standards for light trucks would have a *direct* effect on greenhouse gas emissions; thus, the proposal was a "legally relevant cause" for potential greenhouse warming and required consideration under NEPA. The court noted that the NHTSA did not dispute the fact that fuel economy improvements could have a significant impact on CO_2 emissions, and therefore might have an effect on climatic changes.[59]

Friends of the Earth, Inc. v. Mosbacher[60]   The plaintiff alleged that the Overseas Private Investment Corporation (an export-import bank) failed to comply with NEPA. Specifically, it was alleged that federal funding and loan guarantees for overseas energy projects were provided without a corresponding GHG emission assessment and their potential climatic effects.

In contrast to the case of the *Center for Biological Diversity* (described above), the US Northern District of California noted that oil and gas extraction and power plant projects emit GHGs. The court noted it would be difficult to conclude that there was a genuine dispute that GHGs do not contribute to global warming and suggested that future NEPA climate change litigation could turn on whether a particular agency's action was the "but-for" cause of effects on the domestic environment.

"Remote and highly speculative" impacts

Courts have generally ruled that "remote and highly speculative" impacts that bear only an attenuated relationship to the proposed action need not be analyzed.[61]

Determining when to perform a GHG assessment

Based partly on aforementioned criteria, the author has developed a tool—the Rockville Review Test (see Figure 2.1)—for determining when to perform a NEPA GHG assessment. The tool is so named because the author finalized it while residing near Rockville, Maryland. The significant departure principal (SDP) referred to in Figure 2.1 is described in Chapter 1 of this book as "The Cumulative Impact Paradox."

Best professional practices for performing GHG assessments

Based on existing case law, it appears that global climate change is not an impact that is apart from or can be dismissed from other reasonably foreseeable environmental effects. To the extent that existing scientific evidence provides a basis for evaluating such impacts, analysts are expected to make a reasonable forecast of such impacts.

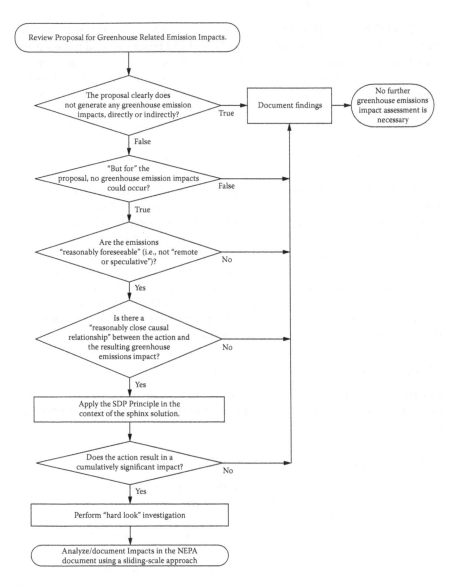

Figure 2.1 Rockville Review Test. Determining whether greenhouse emission impacts must be analyzed in a NEPA document. This analysis uses the significant departure principle (SDP), which is described in an accompanying section.

As of this writing, specific guidance on how to perform an adequate assessment of GHG emissions is limited. This section is designed to provide the reader with some guidance and best professional practices (BPPs) for assessing GHG emissions.

Performing the GHG impact assessment

Traditional techniques, which simply involve reporting a GHG emission such as "GHG emissions would only contribute an increase of 0.0002% to the total annual global emissions," may not be sufficient to address controversial issues such as global warming impacts. Instead, the analysis may need to focus on explaining, perhaps in a macro way, how the proposal would contribute to global GHG concentrations and perhaps affect any trend in such emissions, and then consider how these emissions could affect the environment (e.g., rising sea levels, increased storm activity, severe weather, flooding, or reduced access to water). Moreover, emphasis on potential mitigation measures may be of particular importance. For example, the analysis might focus on best management practices that would conserve energy and reduce GHG emissions. The reader is cautioned to consult with NEPA, regulatory, and legal counsel in determining the most appropriate methods and approaches for fulfilling the intent of NEPA with regard to GHG emission calculations and climate change.

De Minimis actions

Decisions regarding the extent to which an action warrants a NEPA assessment are generally reserved to the expertise and discretion of the agency decision maker as long as it conforms to generally accepted scientific methods and is not perceived to be arbitrary or capricious.

Questions have been raised as to whether a *de minimis* approach could be established for limiting GHG emission analyses. Neither the CEQ nor the NEPA regulations prescribe specific threshold metrics for use in interpreting significance; the courts have generally left such matters to the agency's expertise and discretion. Categorical exclusions might be established for exempting some activities from a GHG analysis.

GHG emissions versus environmental impacts

It is important to emphasize that GHG emissions are not actual impacts in themselves; they are better viewed as the root cause of a change to an environmental resource (impact). One of the important outcomes in *Center for Biological Diversity v. National Highway Traffic Safety Administration* was that the court appeared to be suggesting that simply quantifying emissions and comparing them to a baseline is insufficient;[62] instead, agencies need to actually evaluate how greenhouse emissions affect climate

Table 2.8 Examples of GHG Impacts

Changes in demographics

Effects on agricultural production and food supplies

Temperature variations and their effect on species

More frequent extremes in weather (wetter monsoons and dryer droughts)

Changing precipitation patterns, including droughts and floods

Warmer ocean temperatures affecting weather patterns, coral reefs, fisheries, and tourism

Sea level changes and their effects on coastal zones

Spread of diseases

change. For instance, the actual impacts or consequences are the environmental changes that result from increased GHG concentrations. Examples of GHG impacts are depicted in Table 2.8.

In preparing the analysis, remember that the contribution of the various gases is independent and not additive (Dalton's law of mixed gases).

Standard of care expected in evaluating climate change

The experience of the Gulf Coast during the 2005 hurricane season shows how ill prepared society is for managing weather extremes. The consequences of future climate changes may be measured in billions of dollars and thousands if not millions of lives. Potential effects of climate change are likely to play an increasing role in NEPA analyses, and expectations in terms of what constitutes a benchmark known as standard of care.

The concept of a *reasonable standard of care* is central to understanding future expectations about the quality of the global environment. The concept dates from English common law, which held that activities should be performed in a careful and prudent manner, consistent with the skill level practiced by other members of a given profession. The professional expectation of a standard of care is not the same as legal standards or compliance with regulatory requirements. For most scientific and engineering disciplines, accepted practices that constitute standard of care tend to evolve over time in response to changing conditions such as the degree to which a certain issue can be accurately assessed. The standard is also heavily influenced by technological advancements. With respect to NEPA, planners need to be alert to changes in the standard of care related to the analysis of GHG emissions and climate change. Practitioners can exercise and improve the standard-of-care by reading the professional literature, attending training classes, and participating in professional associations.

Sliding-scale approach

The author recommends the use of a *sliding-scale approach* in determining the scope and level of effort that should be devoted to a GHG emission assessment. With respect to NEPA, a sliding-scale approach[63]

> ... *recognizes that agency proposals can be character-ized as falling somewhere on a continuum with respect to environmental impacts. This approach implements CEQ's instruction that in EISs agencies "focus on sig-nificant environmental issues and alternatives (40 CFR 1502.1) and discuss impacts 'in proportion to their signif-icance' (40 CFR 1502.2[b]). The reader should note that under CEQ's regulations and judicial rulings, a factor in determining significance involves the degree to which environmental effects are likely to be controversial with respect to technical issues.*

Where GHG emissions would be very small, NEPA documents might provide only enough discussion to demonstrate why further analysis is not warranted. Thus, a lower-end sliding-scale approach may only neces-sitate a brief qualitative assessment.

Where potential GHG emissions could be large, an in-depth inves-tigation may be necessary. Such assessments should cite key findings of relevant studies to address potential consequences of GHG emissions (e.g., IPCC assessment reports and other IPCC studies, the US Climate Change Science Program, and studies by other authoritative bodies such as the EPA and National Research Council). An upper-end analy-sis (potentially large-scale impact) may require extensive investigation, complex computer modeling, and a detailed discussion of the results. For evaluations near the upper end of the sliding scale, in addition to performing an explicit analysis of direct, indirect, and cumulative cli-mate change impacts, an analysis may also need to provide estimates of

- Specific changes to global CO_2 concentrations
- Global mean surface temperature and sea level rise
- Rainfall changes
- Loss of species
- Spread of diseases
- Socioeconomic impacts such as dislocations

Analytical considerations

This section provides guidance and speaks to specific analytical consider-ations that must be addressed in a NEPA GHG assessment.

Rule of reason   NEPA imposes strong procedural requirements on agencies to take a "hard look" at the significant impacts of a proposed action. However, this requirement is tempered by the rule of reason. In addressing this balancing act, the US Supreme Court has stated that, consistent with the rule, agencies need to[64]

> *... furnish only such information as appears to be reasonably necessary under the circumstances for evaluation of the project rather than to be so all-encompassing in scope that the task of preparing it would become either fruitless or well nigh impossible.*

Building on this guidance, the CEQ has described the rule as a[65]

> *... judicial device to ensure that common sense and reason are not lost in the rubric of regulation.*

Wishnie describes the scope of federal actions that are subject to NEPA:[66] Only those environmental impacts that bear a "reasonably close" relationship to the major federal action which is the subject of the EIS fall within the reach of NEPA—a relationship akin to the tort doctrine of proximate cause.[67] The length of NEPA's causal chain[68] can be determined by examining the "underlying policies or legislative intent"[69] and considering the legal responsibility of the actors.[70] Impacts that are deemed remote and highly speculative and bear only an attenuated relationship to the proposed action need not be analyzed.[71]

Limitations and problems in evaluating climate change impacts   Analysts should remember that the current state of the art often does not provide a basis for making a precise statement such as "X tons of CO_2 per year would increase the global average temperature of 0.003°C." Although it may be possible to conclude in general terms that climate change will affect individuals, populations, communities, and ecosystems, it can be much more difficult to make meaningful predictions with any specificity as to how those effects will actually manifest themselves. Difficulties in making such projections are related to uncertainty in feedbacks, thresholds, adaptation, location and local variables, resilience and component interactions, as well as substantial uncertainty inherent in the models used to develop climate projections and the degree of natural spatial and temporal variability in climate parameter characteristics of a region.

Nonlinearity and tipping point   Changes in climatic conditions may result in complex and unpredictable interactions. Practitioners may want to note in the analysis that a continued increase in GHG emissions might involve a *sudden and nonlinear* change in climate; this potential phenomenon has been referred to as the tipping point and such events are not well understood. Climate change feedback loops can be positive (increasing) or negative (decreasing or ameliorating the effects). Consider the Arctic environment. Some feedback loops are well understood. For instance, warmer temperatures result in reduced sea ice extent, causing less solar heat to be reflected back to space by the ice cover, and more to be absorbed by the sea, resulting in increased rates of melting, which further reduce sea ice cover, resulting in more warming. Positive feedback may come from the release of GHGs (CO_2 and methane) from thawing permafrost and oceanic warming. Negative feedback loops may result from changes in salinity produced by increased freshwater delivery to the ocean from ice melt and river discharge, causing changes in ocean circulations and reducing heat transported to the Arctic.

As a second example, consider the fact that increased insect populations resulting from milder winters could promote insect disease vectors; however, weather patterns may also result in changes in wind patterns, thereby reducing insect migration. Earlier ice melting might increase nesting areas, yet change conditions that may reduce over-winter survival. Vegetation change may eliminate or reduce reproductive areas in one place but enhance them in other areas. Because not all species will react to changing conditions in the same manner, some species populations may actually increase, while others may decline (or actually increase), and still others may deteriorate toward extinction.

Thresholds are levels that once crossed may result in changes disproportionate to the increment that caused the threshold to be crossed. A change in climatic conditions should not be assumed to occur in a linear fashion. Abrupt changes may occur as thresholds are breached. Unfortunately, many thresholds are difficult to quantify; this problem is compounded for thresholds on which there is little scientific data, and for those in which there can be a long time lag between a breached threshold and an observable response.

Mitigation measures and analysis of a carbon-neutral program
One approach is to focus alternatives and mitigation measures on a carbon-neutral program. In this case, the analysis should take credit for activities that can offset the GHG impacts:

- Environmental awareness programs
- Recycling
- Carbon sequestering (if practical)
- Mulch programs

For example, recycling 1,000 pounds of paper as opposed to manufacturing it from virgin materials can save

- 15 trees; the 15 saved trees can absorb between 120 and 220 pounds of CO_2 each year (Burning this paper would *create* carbon emissions.)
- 750 to 1400 pounds of CO_2
- 150 gallons of oil
- 2,000 kilowatt-hours of energy
- 4,000 gallons of water

This represents a 60% savings in energy (which may mean less GHG emitted from gas- or coal-fired plants).

Addressing indirect effects of GHG emissions

Based on earlier discussions of these considerations, the following sections provide practical guidance for performing a GHG emission and climate change assessment. This section summarizes a paper by Moore et al., which investigated requirements for analyzing indirect effects of GHG emissions in NEPA documents.[72] *Mid-States Coalition for Progress v. Surface Transportation Board* involved an example of an indirect GHG impact. Potential GHG emissions from Midwestern power plants were found to be an indirect effect of the board's decision to license a rail line that would increase the volume of coal available to the market.[73] Based partly on this case, the following guidance is suggested for addressing indirect effects of greenhouse emissions in NEPA documents.

The rule of reason and reasonably foreseeable standards

Consistent with the rule of reason, an indirect effect considered in an NEPA analysis should meet the criteria of being "reasonably foreseeable" (40 CFR §1508.8). Thus, an effect is considered reasonably foreseeable if it is[74]

> *sufficiently likely that a person of ordinary prudence would take it into account in reaching a decision.*

Providing additional clarification, the Ninth Circuit employed a "links of a chain" analogy, that is, how far removed down a chain of events can an indirect impact be taken before it is no longer subject to NEPA. For instance, a proposal to construct a project that would require a supply of steam so large that an adjacent private steam vendor would have to build an additional coal-fired steam plant to produce that steam is likely to constitute a causal chain of events that are reasonably foreseeable; the indirect impact (a new coal-fired steam plant) would also have to be addressed. In contrast, a proposal to develop mine in New Mexico that would produce iron which would eventually be forged into steel in multiple (but unknown) plants in Michigan, and require burning more coal to produce that steel may well constitute an excessively long causal chain of events that are not reasonably foreseeable. The courts continue to struggle with determining a definable extent to which the causal chain may be taken.

More to the point, indirect impacts should be discussed in proportion to their *proximity* to the precipitating action. This is a reasonable application of the rule of reason. A GHG emission may be significant; but if the only causal link between the GHG emitting project and a resulting change in global climate is through a long chain of events (construction, operational, and possibly increased traffic resulting in marginal GHG emissions), a detailed discussion may be disproportionate given this chain of events. In considering factors such as proximity, analysts may find it unnecessary to describe potential global warming changes that constitute a *very indirect effect*, compared with the direct effects of the proposal, which might require a much more rigorous assessment. Indirect impacts analyses should normally be addressed in relation to their proximity to the action, allowing an agency to disclose (but not excessively evaluate) some highly attenuated impacts without fear of having the adequacy of those disclosures challenged.

Causation and remoteness

In one Supreme Court case, plaintiffs asserted that the Department of Transportation (DOT) violated NEPA by not considering potential effects of the increase in cross-border operations of Mexican motor carriers resulting from the DOT's propagation of regulations related to the President's lifting of a moratorium on Mexican motor carrier certifications.[75] Plaintiffs maintained that the increased truck emissions were reasonably foreseeable indirect impacts of the regulations. The Supreme Court concluded that "NEPA requires a 'reasonably close causal relationship' akin to proximate cause in tort law;" because DOT could not prevent the cross-border operations of Mexican motor carriers, the impacts of such operations would not be subject to DOT's decision-making process. While the "but-for" criteria may be sufficient to trigger a GHG assessment, they may not always be sufficient to require an analysis of indirect impacts.

As described in "Criteria Useful in Determining the Need to Evaluate," courts have tended to place great reliance on consideration of "reasonably foreseeable" impacts to exclude more speculative and remote effects. They also consider the questions of causation. In one case, the Corps of Engineers issued a permit for a riverboat gambling facility. The court clearly separated the direct and indirect impacts of the action. Here, direct permitting effects included dredging the river; indirect impacts included construction of a hotel, golf course, and parking facilities. The court concluded that such development was not reasonably foreseeable and the Corps' decision to exclude them from the analysis was consistent with the rule of reason. The *Hoosier* court distinguished this case from an earlier *Davis* decision by the Ninth Circuit Court, concluding that the two cases were distinctly different. In *Davis*, development would definitely result from a proposal to build a major interchange in an agricultural area. In *Hoosier*, it could not be definitely shown that the proposal would induce a reasonable person to foresee the commercial development that might result from the presence of the riverboat casino.

Yet, several courts have struck down agency decisions for failing to adequately evaluate the growth-inducing effects of major federal projects, especially where the goal and anticipated result of the project were to stimulate growth and development. These cases show that agencies face substantial uncertainty about the extent to which their NEPA analyses must consider more and more remote (indirect) impacts of an action.

Method for evaluating climate change impacts

The courts have ruled that US federal agencies must take a "hard look" at environmental impacts of their activities. This section describes a systematic process for evaluating GHG impacts. The first subsection describes a minimum five-step procedure for assessing potentially significant GHG impacts. The second subsection describes a 15-step method for evaluating climate change impacts; it should be followed in preparing more extensive GHG impact assessments, where the GHG emissions or their potential impacts are particularly significant.

Minimum five-step procedure for assessing GHG impacts

Table 2.9 depicts a minimum five-step procedure for assessing potentially significant GHG impacts.

Table 2.9 Minimum Five-Step Procedure for Assessing GHG Impacts

Identify and quantify the amounts of each GHG emission (and as appropriate, provide a total in CO_2 equivalents); be conscious of the fact that certain gases such as methane are significantly more potent GHGs than CO_2.

Investigate potential means for avoiding GHG emissions. As reasonable, include alternatives for reducing emissions; if no reasonable alternatives are available, the NEPA document should state this fact.

Identify and investigate reasonable mitigation measures for minimizing or compensating for GHG emissions.

Document the assumptions and scientific methods used in analyzing the impacts.

Analyze the impacts of these GHG emissions (or reductions or offsets) based on best existing data (noting incomplete or unavailable data per 40 CFR §1502.22).

Note: see Table 2.11.

A fifteen-step method for preparing a comprehensive GHG assessment

Bass has recommended a 10-step approach for evaluating GHG emissions and climate change impacts in NEPA documents.[76] The author proposes a modified 15-step, general-purpose method for evaluating GHG emission impacts (Table 2.10).

Examples of describing GHG impacts in NEPA documents

Emissions are typically presented as annual rates. Other analytical considerations may include total life-cycle emissions, and the potential to trigger other related actions, which produce additional emissions.

Where the effects of a proposal contributing to GHG emissions have been considered, federal agencies have typically discussed the issue in the air impacts section of the NEPA document. However, the issue can also cut broadly across many resource disciplines. For instance, from a national forest planning context, one agency considered impacts of global climate change on forest hydrology, insects and pathogens, vegetation, wildlife, fire regimes, and air quality within the physical and biological environment sections of the EIS.[77] Such issues should also be addressed in the cumulative impacts section.[78] In some instances, agencies have used various modeling techniques to investigate potential additional GHG emissions, and compared that increase to global or United States total emissions.[79]

The analysis should also describe how emissions will relate to achievement of agency GHG reduction goals required under E.O. 13514 and to achievement of state GHG reduction goals and laws. Such compliance is also a factor in determining significance (40 CFR 1508.27[b][10]).

How GHG emissions were addressed in two EISs

Described below are examples of how two recent US Department of Energy (DOE) EISs addressed GHG emissions.

Futuregen project EIS This EIS addressed climate change impacts using statements and evidence such as[80]

- While "CO_2 is not currently regulated as an air pollutant at the federal level, it is generally regarded by a large body of scientific experts as contributing to global warming and climate change (IPCC, 2007)."
- The EIS analyzed a coal-fueled electric power and hydrogen production plant integrated with CO_2 capture and geologic sequestration. Such a design would be capable of capturing at least 90% of its CO_2 output.
- The project's individual contribution to global CO_2 emissions and potential climate change is extremely small.

Table 2.10 Fifteen-Step General-Purpose Method for Preparing
Comprehensive GHG Assessment

Scoping:
1. Define the temporal and spatial context of the climate change analysis. Note: in some cases that this context may not be local or regional, but global in extent.
2. Consider the direction provided in Sections 2.5 and 2.6. Determine if there is a causality between the action and potential climate change effects. With the guidance of regulatory or legal counsel, consider the following criteria:
    (a) "But for" the proposed project: Identify related or indirect actions (which could also affect climate change) that would not occur "but for" the implementation of the proposal; or
    (b) The extent to which such effects would be *caused* by the proposed federal action.
3. Identify potentially significant emissions and/or climate change issues that the analysis will focus on.
4. Identify potential reasonable alternatives for avoiding or reducing emissions.
5. Identify potential mitigation measures for reducing emissions or compensating for their impacts.

Affected environment:
6. Describe the temporal and spatial bounds of the analysis.
7. Describe the potentially affected environment that might be affected by potential changes in climate.
8. Describe the global GHG inventory.
9. Describe other applicable local, state, or national laws or regulations that may be related to this analysis.

Alternatives:
10. Describe any reasonable alternatives or mitigation measures that will be investigated in the analysis for avoiding, reducing, or mitigating GHG emissions and potential climate change impacts.

Environmental impact assessment:
11. As practical, quantify the project's direct and indirect GHG contributions or emission offsets. If this is not possible or is impractical, provide a qualitative description of these emissions. As practical, use validated modeling software to compute the effect of the GHG contributions, offsets, and corresponding impacts.
12. As practical, quantitatively describe (or provide a qualitative impact if a quantitative assessment is impractical) the cumulative emissions and climate change impact. Describe how this impact could affect ecosystems and socioeconomic attributes of human society.
13. Investigate and describe the effectiveness of reasonable mitigation measures for reducing GHG emissions and impacts.

Table 2.10 Fifteen-Step General-Purpose Method for Preparing
Comprehensive GHG Assessment (Continued)

14. Using regulatory direction provided in the NEPA Regulations (40 CFR
1502.22), describe any uncertainties, including incomplete or unavailable
information (see Table 2.11).
15. As applicable, develop a monitoring plan, possibly in conjunction with an
environmental management system or adaptive management process to
monitor GHG emissions.

Note: see Table 2.11.

The Gilberton coal-to-clean fuels and power project EIS   This EIS included
statements and evidence such as[81]

- The CO_2 emissions from the proposed facility would add 2.3 million
  tons per year to global CO_2 emissions, for an estimated cumulative
  increase of 29 billion tons.
- "Fossil fuel burning is the primary contributor to increasing concentrations of CO_2 …. The increasing CO_2 concentrations likely have
  contributed to a corresponding increase in temperature in the lower
  atmosphere."
- "Over the entire fuel lifecycle (from production of the raw material
  in a coal mine or oil well through utilization of the fuel in a vehicle)
  and considering all greenhouse gases, production and delivery of
  liquid transportation fuels from coal has been estimated to result in
  about 80% more greenhouse gas emissions than from the production
  and delivery of conventional petroleum-derived fuels…. Recovery
  and sequestration of CO_2 at a … production facility … could greatly
  reduce greenhouse gas emissions from … fuel production, possibly
  to levels below conventional petroleum-derived fuel production."
- "Although not proposed by the applicant, it may become feasible to reduce the project's contribution to global climate
  change by sequestering some of the CO_2 captured in the process
  underground."
- "Using high-range estimates of future oil prices … and assuming
  the … fuel cycle generates 80% more greenhouse-gas emissions than
  production and delivery of conventional petroleum-derived fuels,
  expanded use of [this] technology to produce liquid fuels could
  cause the US liquid fuel sector to release about 5% more greenhouse
  gas emissions than if the same quantity of fuel was produced from
  petroleum."

Dealing with uncertainties, including incomplete or unavailable information

Research on climate change impacts is an emerging and rapidly evolving area of science. Given the state-of-the-art, it may be impossible for an agency to make definitive statements concerning the impacts of GHG emissions. The NEPA regulations provide for instances where an analysis of an impact lies beyond the state of the art or involves incomplete or unavailable information. The regulations emphasize that "… when the nature of an effect is reasonably foreseeable but its extent is not, the agency cannot simply ignore the effect" (40 CFR 1502.22). If the NEPA analysis involves uncertainties, such as incomplete or unavailable information, relevant to reasonably foreseeable significant adverse impacts is essential to a reasoned choice among alternatives; if the overall costs of obtaining it are not exorbitant, the agency shall obtain and include the information in the NEPA analysis.

However, when a NEPA analysis involves uncertainties (including incomplete or unavailable information) that can affect the evaluation of reasonably foreseeable significant adverse effects, it must comply with provisions provided in 40 CFR §1502.22 of the NEPA regulations. This direction is applicable to the preparation of both EISs and EAs. Table 2.11 provides guidance for complying with this provision.

For instance, the recent NHTSA draft EIS on proposed new corporate average fuel economy (CAFE) standards for passenger cars and light trucks includes a substantial discussion of GHG emissions in response to a 2007 court order. Because the proposal involves substantial uncertainty, including incomplete or unavailable information regarding the potential impacts, one of the statements presented in the draft EIS reads as follows:

> *…the magnitudes of the changes in these climate effects that the alternatives produce—a few parts per million (ppm) of CO_2, a hundredth of a degree C [centigrade] difference in temperature, a small percentage-wise change in the rate of precipitation increase, and a 1 or 2 millimeter… sea level change—are too small to meaningfully address quantitatively in terms of their impacts on resources. Given the enormous resource values at stake, these distinctions may be important—very small percentages of huge numbers can still yield substantial results—but they are too small for current quantitative techniques to resolve….*

Table 2.11 Direction for Dealing with Uncertainties, Including Incomplete or Unavailable Information (40 CFR §15022)

When an agency is evaluating reasonably foreseeable significant adverse effects on the human environment in an environmental impact statement and there is incomplete or unavailable information, the agency shall always make clear that such information is lacking.

(a) If the incomplete information relevant to reasonably foreseeable significant adverse impacts is essential to a reasoned choice among alternatives and the overall costs of obtaining it are not exorbitant, the agency shall include the information in the environmental impact statement.

(b) If the information relevant to reasonably foreseeable significant adverse impacts cannot be obtained because the overall costs of obtaining it are exorbitant or the means to obtain it are not known, the agency shall include within the environmental impact statement:

1. A statement that such information is incomplete or unavailable

2. A statement of the relevance of the incomplete or unavailable information to evaluating reasonably foreseeable significant adverse impacts on the human environment

3. A summary of existing credible scientific evidence that is relevant to evaluating the reasonably foreseeable significant adverse impacts on the human environment

4. The agency's evaluation of such impacts based upon theoretical approaches or research methods generally accepted in the scientific community. For the purposes of this section, "reasonably foreseeable" includes impacts that have catastrophic consequences, even if their probability of occurrence is low, provided that the analysis of the impacts is supported by credible scientific evidence, is not based on pure conjecture, and is within the rule of reason.

The questions, issues, and complexities faced in preparing a GHG assessment

An assessment of climate change, in particular, can pose difficult and unique analytical challenges. For example,

- When assessing transportation emissions for a road construction project, how should the NEPA assessment consider the fact that if the project were not built, drivers might simply travel elsewhere instead, raising doubt about the net increase in transportation emissions?
- How can practitioners provide a sufficient description of the environmental "baseline" (e.g., current and future GHG concentration

levels, encroachment of non-native species, or a decreasing ice shelf) if the baseline is evolving over time or in unpredictable ways?
• How does an agency perform a reasonable and meaningful cumulative effects analysis of climate change?

Among ambiguous NEPA legal issues are questions relating to the

1. Applicability of NEPA to federal agency actions supporting overseas projects, that emit GHGs, which may impact the domestic US environment[82]
2. Degree to which a NEPA document must consider secondary impacts, such as global warming impacts that might result from increased use of coal if a new rail line were approved to transport coal to another region of the United States[83]

Some additional timely and important questions and issues are listed in Table 2.12.

Considering cumulative GHG impacts

Assessing the significance of an environmental impact, pursuant to the NEPA, requires consideration of direct, indirect, and cumulative

Table 2.12 Some Important Unsettled GHG Assessment Issues

To what extent (if any) should GHG emissions beyond the borders of the United States be considered?
What specific variations in climate changes should be assumed? For example, should a worst-case sea level scenario be assumed; should severe hurricanes be assumed?
Should potential climate change impacts alone be sufficient to trigger the need for an EIS, or must some other significance criterion also be trigged? Will the EIS be treated only as a disclosure document, or should increased emphasis be placed on adopting reasonably identified mitigation measures?
Should greater emphasis be placed on adopting reasonable alternatives, such as smaller or less GHG producing projects if they are shown to have less effect on climate?
How far upstream should an analysis be carried out? For instance, should a coal-fired power plant project consider the impacts of coal mining?
Should GHG emission guidelines be adopted for various kinds of projects?
Should a climate change analysis be required for all projects subject to a NEPA review or only those of a certain type or over a certain size?
If mitigation is adopted, how will compliance be enforced and monitored?
Are carbon offset purchases or carbon trading considered acceptable mitigation measures? Can such offsets be purchased from anywhere, or are they restricted to certain geographic or national areas?

impacts.[84] Thus, decision makers must consider cumulative impacts, along with direct and indirect impacts, in reaching a decision regarding the significance of a proposed action. This fact is underscored by the following NEPA regulatory provisions. Specifically, a "cumulative impact" is defined as an environmental impact that results from[85]

> *... the incremental impact of the action when added to other past, present, and reasonably foreseeable future actions, regardless of what agency (federal or non-federal) or person undertakes such other actions. Cumulative impacts can result from individually minor but collectively significant actions taking place over a period of time.*

The CEQ handbook entitled *Considering Cumulative Effects under the National Environmental Policy Act* acknowledges three basic types of cumulative effects:[86]

1. *Additive* (loss of sensitive resources from more than one incident)
2. *Countervailing* (negative effects are compensated for by beneficial effects)
3. *Synergistic* (total effect is greater than the sum of the effects taken independently)

Additive impacts are often difficult to detect and do not necessarily add up in the strict sense of 1 + 1 equaling 2; it is possible that an additive effect of two equal environmental disturbances could be greater than one but less than two; the converse is also true. A *synergistic effect* is a total effect that is greater than the sum of the additive effects on a resource; for example (continuing with the numerical analogy), the total cumulative effect from two equal inputs would be greater than two. A *countervailing impact* is one that subtracts or compensates for other impacts, such as a carbon sink that reduces total CO_2 emissions.

Depending on the nature of the proposal and the amount of potential GHG emissions, a cumulative impact assessment (CIA) might discuss the following conceptual elements:

- Potential to spawn other actions.
- Combination with other emissions. For example, "The proposed facility would add X kilograms per year of CO_2 to existing (or projected future) emissions of Y kilograms per year from fossil-fuel combustion, and Z kilograms from all other sources."
- Total emissions over the project lifetime. As appropriate, the assessment may require a life-cycle analysis.

Private projects as major federal actions under NEPA

Determining when an essentially non-federal project that produces GHG emissions becomes "federalized" to such an extent that it triggers NEPA can be complicated. To date, such questions have centered on federal financing of projects.

The case of *Friends of the Earth, Inc. v. Mosbacher*[87] provides some insight. Here the court concluded that significant federal funding could be sufficient to trigger a project into a major federal action for the purposes of NEPA, but that both the nature of the federal funds and the extent of federal involvement must also be considered. Because no established standards exist for defining when partial federal financing transforms a project into a major federal action, the court considered the extent to which "an agency that provides financing to a project can influence or possess actual power to control a non-federal activity."

In this case, the court concluded that there was insufficient information in the record to decide whether the agency's involvement rose to a level that would constitute a major federal action, or whether the defendants would be the relevant cause of alleged climatic effects within the United States. The case has since been turned back to the District Court in San Francisco.

A complaint was also recently filed involving *Montana Environmental Information Center v. Johanns.*[88] This case alleges a failure to assess the cumulative impacts of federally financed projects on global warming.

Spatial and geographic considerations

GHG emissions and their potential effects on global climate know no bounds. NEPA was designed to address geographically bounded environmental concerns—not global problems such as climate change. The geographic requirements of the analysis are particularly important in performing such analyses. The principal responsibility for determining the geographic scope of a cumulative impact analysis lies with the lead agency responsible for preparing the EIS.[89]

As traditionally viewed by the courts, projects included in a cumulative impact analysis have typically fallen within the same geographic area.[90] With regard to causal relationship, most courts have generally reaffirmed earlier rulings that all reasonably foreseeable actions must be analyzed in the cumulative impact context.[91] With respect to the issue of global climate change, it remains unclear if and to what extent international impacts of US projects and the US impacts of actions by international actors should be treated in a cumulative impact analysis.

Assessing cumulative significance

Most CIAs have some type of natural or physical boundary. NEPA analysts typically use these natural or physical boundaries to define the geographical extent of a cumulative impacts analysis. Consider a proposal to build a power plant that produces sulfur dioxide. The NEPA analysis will typically examine the impacts of such emissions within the local air-shed. The analysis will also evaluate other "reasonably foreseeable future actions" that would also generate sulfur dioxide within the airshed. The analysis will also determine the time frame within which this gas can be expected to affect the airshed.

However, the extents of CO_2 and many other discharged GHGs are distinctly different from most other environmental disturbances in that their context is global. Thus, there is no geographically limited boundary for such an analysis. GHG emissions cross international borders and, consequently, a CIA analysis that does not account for the international context will fail to fully account for the global impact of GHG emissions. With respect to GHG analysis, the airshed could be considered to be the entire world. However, many GHG emissions can be expected to remain in the atmosphere for a century or more.

As stated in 40 CFR §1508.27(a), significance can vary with the context or degree of area impacted. As just witnessed, the impact of most projects is typically limited in geographic scale. A finding of significance is therefore frequently avoided by shifting the scale of the affected environment (site-specific versus region or national impact); similarly, a finding of significance has often been avoided because the proposal only impacts an individual or a small local population or species versus the environment as a whole.

As noted earlier, the US Court of Appeals for the Ninth Circuit recently found the NHTSA's EA for CAFE standards for light trucks inadequate in several respects, including the analysis of cumulative impacts. The court stated that

> *Any given rule setting a CAFÉ standard might have an 'individually minor' effect on the environment, but these rules are 'collectively significant actions taking place over a period of time.*

The cumulative impact paradox

Over the past century and a half, emissions from past and present actions have raised global CO_2 concentration emissions about 25%. Many climatologists, as well as the 2007 IPCC report, have concluded that this increased CO_2 concentration is probably already adversely affecting the Earth's climate and natural resources.

A strict regulatory interpretation of significance can also lead to a paradox (Eccleston's cumulative impact paradox) when one considers how the environment appeared before the intervention of human activities. A

FONSI, by its very definition, states that an action will not have a significant effect, including a cumulatively significant effect. Because the global GHG emission concentration has already breached a level that most decision makers consider significant, this Paradox is particularly problematic in assessing proposals that emit even innocuous levels of GHGs.

Because the global GHG concentration is generally considered to have already exceeded a significant concentration (i.e., cumulatively significant impact), a logical paradox arises in which many, if not most, federal activities should require preparation of an EIS as they are technically ineligible for a FONSI (as described shortly, a parallel problem is also encountered in applying categorical exclusions and assessing what would otherwise be deemed to be non-significant impacts in an EIS). A strict interpretation of significance leads to such a conclusion, even in cases where the direct and indirect GHG impacts of a proposed activity are finite but essentially innocuous. As described in the following section, NEPA practitioners and decision makers are slowly beginning to appreciate the implications of this paradox (described in the next section). This paradox must be resolved, if the analysis of cumulative GHG emissions is to contribute in a meaningful way to federal decision-making. The author refers to the approach for resolving this paradox as the sphinx solution because he finalized the concept while touring the Great Sphinx in Egypt. This approach expands upon the *significant departure principal* (or sometimes referred to as the significant difference principal, see Section 1.7) presented in Chapter 1. This concept builds on a publication published in a 2006 issue of the *Journal of Environmental Practice.**

The paradox and the greenhouse assessment problem

Assessing the significance of an environmental impact, pursuant to the NEPA, requires consideration of direct, indirect, and cumulative impacts [40 CFR 1508.8(a) and (b), 1508.25(c), 1508.27(b)(7); see Council on Environmental Quality (CEQ), 2005]. Thus, decision makers must consider cumulative impacts, along with direct and indirect impacts, in reaching a decision regarding the significance of a proposed action. This fact is underscored by the following NEPA regulatory provisions. Specifically, a *cumulative impact* is defined as an environmental impact that results from

> the incremental impact of the action when added to other past, present, and reasonably foreseeable future actions, regardless of what agency (federal or non-federal) or person undertakes such other actions. Cumulative impacts can result from individually minor but collectively significant actions taking place over a period of time. (40 CFR 1508.7.)

* Eccleston, C. H., Applying the significant departure principle in resolving the cumulative impact paradox. *Journal of Environmental Practice*, 8(4), December 2006.

The NEPA implementing regulations (hereafter referred to as the regulations) go on to state:

> *Significance exists if it is reasonable to anticipate a cumulatively significant impact on the environment. Significance cannot be avoided by terming an action temporary or by breaking it down into small component parts. (40 CFR 1508.27(b)(7).)*

The importance of assessing cumulative impacts is underscored by one of the factors provided in the regulations, which is required to be considered in reaching a determination of significance:

> *Whether the action is related to other actions with individually insignificant but cumulatively significant impacts. Significance exists if it is reasonable to anticipate a cumulatively significant impact on the environment. Significance cannot be avoided by terming an action temporary or by breaking it down into small component parts. (40 CFR 1508.27(b)(7).*

Assessing cumulative significance can lead to a paradox

Now consider that the U.S. Council on Environmental Quality (CEQ) estimates that between 30,000 and 50,000 NEPA EAs are prepared each year. A FONSI for an EA, by definition, means that an action

> *will not have a significant effect on the human environment. (40 CFR 1508.13).*

As described in the previous section, the phrase "will not have a significant effect" implies that the action will not contribute to a cumulatively significant impact. Moreover, A categorical exclusion (CATX) is defined to mean:

> *... a category of actions which do not individually or cumulatively have a significant effect on the human environment and which have been found to have no such effect in procedures adopted by a Federal agency in implementation of these regulations (Sec. 1507.3) and for which, therefore, neither an environmental assessment nor an environmental impact statement is required.*

Thus, by these NEPA regulatory provisions, neither a CATX nor FONSI can be issued for any action that results in or contributes to a cumulatively

significant impact. Now, if an environmental resource has already sustained a *cumulatively significant* impact, how can a decision-maker declare
that a proposed action is eligible for a CATX or FONSI? A strict interpretation of the Regulations leads to the conclusion that a CATX or FONSI
can be issued only if the proposed action's contribution to the cumulative
impact is zero. Thus, with respect to NEPA, the cumulative impact paradox can be summarized from three different perspectives:

> A strict interpretation of cumulative significance leads to the conclu-
> sion that any contribution in a GHG gas emission is cumulatively
> significant regardless of how small the contribution is, preventing
> the application of a FONSI for such an action.
> Similarly, a strict interpretation, also leads to the conclusion that a
> CATX cannot be applied to any action that individually or cumu-
> latively [has] a significant effect. Thus any action is ineligible for a
> CATX if it contributes any amount of GHG emission regardless of
> how small that contribution is.
> The NEPA implementing regulations state that an EIS is to focus on sig-
> nificant impacts. Any impact that is reviewed in an EIS and which
> would normally be determined to be nonsignificant, must be instead
> evaluated as a significant impact if it contributes any amount of GHG
> emission, regardless of how small that contribution would be.

It is reasonable to ask why this situation is considered to constitute a
paradox. A paradox can be defined as a contradiction, absurdity, illogicality, or irony. The paradox stems from at least two perspectives. The
principal purpose of a CATX or FONSI is to demonstrate that the impacts
of a proposal are nonsignificant and therefore an EIS does not need to be
prepared. Yet, a strict interpretation of cumulatively GHG emissions leads
to the conclusion that any amount of emission, regardless of how small,
is cumulatively significant (even where the direct and indirect emissions
are clearly nonsignificant), which would leads to a conclusion that an EIS
must be prepared for such an action. Yet preparing an EIS on every action
that produces even the most innocuous amount of GHG emission would
serve little or no useful purpose. Such a situation certainly meets the criteria of an absurdity, illogicality" or irony.

Secondly, a paradox also results from another, albeit, related perspective. The NEPA implementing regulations place strong emphasis stream-
lining the NEPA process by reducing delays and excessive paperwork.
Consistent with this direction, the purpose for introducing the CATX and
FONSI was to reduce delays and excessive paperwork such that agencies
could focus resources (i.e., preparation of EISs) on environmental issues
that were of real concern (i.e., "truly significant"). Yet a strict interpretation
of cumulative significance leads to the conclusion that perhaps a majority

of CATXs and FONSIs could not be issued. Such a scenario leads to a contradiction with CEQ's original mandate to reduce delays and excessive paperwork (i.e., a contradictory paradox).

Taken to its logical conclusion, this paradox can lead to almost ridiculous deductions. For instance, humans inhale air (largely oxygen and nitrogen) and exhale CO_2 (along with an assortment of other gases). Thus, even an independent proposed Federal action that would only involve a single individual (no vehicles, machinery, or other equipment) would contribute to a cumulatively significant global CO_2 concentration (albeit infinitesimally small increase), thus requiring preparation of an EIS for a single Federal action involving one sole person.

This being the case, any action that results in a net GHG incremental emission would need to comply with one of the three courses of action shown in Table 2.13. More to the point, many (if not most) activities for which EAs are currently being prepared should actually be ineligible for a FONSI and therefore require preparation of an EIS.

The first item in Table 2.13 would lead to the additional preparation of tens of thousands of EISs each year. This would likely spell the end of the NEPA. The second item requires reducing *all* GHG emissions to zero, which would not only be prohibitively costly, but probably represents a level of technology beyond the current state of the art; this, too, would probably the represent the nail in NEPA's coffin. The third item would bring to a halt tens of thousands of projects, many of which are vital and beneficial to our society; again, this option could lay the wreath on NEPA's headstone.

Invalidating the use of many, if not most, FONSIs would lead to preparation of perhaps tens of thousands of additional EISs per year; invalidating the application of many, if not most, CATXs could conceivable lead to the additional preparation of perhaps hundreds of thousands of additional EIS per year. Clearly, the strict interpretation of these regulatory provisions leads to an absurd and unreasonable quandary (Eccleston's paradox) in which tens, perhaps even hundreds, of thousands of EISs might be required for relatively innocuous projects. It is not difficult to understand that the public, politicians, and most pragmatic practitioners would find such conclusion to be unreasonable, impractical, and politically unacceptable.

Table 2.13 Possible Outcomes of Using a Standard Regulatory Interpretation, the Cumulative Impact Paradox

Prepare an environmental impact statement (EIS) to evaluate the impacts, including cumulative incremental greenhouse gas (GHG) emissions.
Avoid or offset the emissions mitigation produced by the proposal (to the point of a zero net increase in carbon emissions) so that the action can qualify for a Finding of No Significant Impact (FONSI).
Do not pursue the action.

A growing consensus on the paradox

Over the past century and a half, emissions from past and present actions have raised global carbon dioxide (CO_2) concentrations by approximately 25%. Many climatologists, as well as the 2007 report by the International Panel on Climate Change, have concluded that this increased CO_2 concentration is probably already affecting the Earth's climate and, as a result, biological and natural environmental resources.

Because the global concentration level has already exceeded a reasonably defined significance threshold, many NEPA specialists are left to conclude that any incremental addition of CO_2 is likewise contributing to this significant impact. It is difficult to justify how a FONSI can be issued to any project that emits any amount of CO_2.

This opinion was also voiced during a panel discussion of lawyers at the 2009 NEPA conference co-sponsored by the Environmental Law Institute, George Washington University Law School, and the CEQ (CEQ and Environmental Law Institute, 2009). Figure 2.2 was presented by one of the lawyers on this panel. This figure shows a bell-shaped curve in which the cumulative CO_2 level has gradually increased to the point where it has breached the GHG significant concentration level (presumably this has already occurred); because this significance level has already been breached, he argued, a strict interpretation of cumulative significance dictates that the CO_2 emission of any project would have to be mitigated to zero to qualify for a FONSI. This requirement would apply to all projects qualifying for a FONSI until the global CO_2 concentration has been reduced to a point below the threshold of significance (trailing end of the bell-shaped curve).

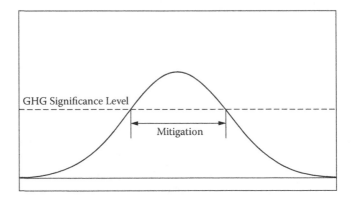

Figure 2.2 A strict interpretation of significance implies that after the global greenhouse gas (GHG) significance level has been breached, mitigation is required to reach a finding of no significant impact (FONSI) for any future action contributing any incremental GHG emission.

Wishnie[92] echoes a similar conclusion:

> *Even the most miniscule emission becomes significant when added to the cumulative impact of global emissions. Retaining the cumulative impacts requirement would create unworkable burden for agencies, extending the NEPA analysis requirement ever-downward towards actions with smaller and smaller impacts. Eliminating the cumulative impacts analysis requirement in conjunction with establishing a significance standard ... is a vital first step towards making GHG emissions impacts analysis feasible.*

Reinke acknowledges the same quandary, arguing that the regulations do not allow for a FONSI in cases involving an incremental emission (even for very small contributions) that contributes to a cumulatively significant impact. He concludes that, from a technical perspective, any project resulting in a net GHG emission increase would need to pursue one of the three actions presented in Table 2.13.

Thus, Reinke argues that under the regulations, "negligible impacts" or "infinitesimal small impacts" are equivalent to a small incremental cumulative impact. Moreover, the regulations fail to define whether it is possible to reach a determination of nonsignificance where an incremental increase contributes to a significant problem. He states that no court has yet specifically addressed this quandary. Reinke views this dilemma as so unworkable as to necessitate changing the regulations to revise the cumulative impact requirement or establish new significance standards. The author also held extensive dialogs with a prominent NEPA attorney who espoused a similar opinion.

In another example, a law professor has gone so far as to propose that either NEPA or its implementing regulations be amended to provide language that avoids a finding of significance for actions based on production of GHG emissions.

This question was also continually raised at the annual conference of the National Association of Environmental Professionals, which was held April 27–30, 2010, in Atlanta, Georgia. When questioned, Lucinda Swartz, former deputy general counsel for the CEQ, concurred that in her opinion, a strict interpretation of cumulative significance leads to the conclusion that any incremental CO_2 emission can be considered to constitute a significant impact requiring preparation of an EIS.[93]

One solution that has been proposed involves preparing numerous programmatic EISs that would assess GHG emissions for all projects or at least a wide assortment of projects. Such an idea has some advantages, as well as obvious problems. Such programmatic EISs might not be required under existing case law. Beyond the obvious reluctance on the part of many agency officials and some members of the public, such a proposition would also require large sums of money and years to complete. But perhaps most troubling is that this solution does not necessarily resolve the regulatory conflict problem. FONSIs and CATXs would still need to be issued to evaluate site-specific details of various actions. But these FONSIs and CATXs would still encounter the same cumulative impact conflicts.

An alternative method that avoids this quandary is provided in the following sections.

Toward a solution

With respect to cumulative impacts, the concept of cumulative significance necessitates that it be interpreted differently from the way it is routinely considered in the assessment of direct and indirect effects. The assessment of direct and indirect effects provides a simpler, independent, and stand-alone analysis of the impacts of a proposal (emission, effluent, noise, scouring) on the baseline environment. Because the assessment of significance for direct and indirect impacts is unconcerned with any other impacts (past, present, or reasonably foreseeable), it is restricted to simply determining whether the effect would produce some important (significant) change in an environmental resource. That is, it is unconcerned with whether the resource has already been or will be significantly impacted. Such an analysis is relatively straightforward because the assessment of direct and indirect impacts focuses only on how an action changes, an environmental resource.

But this is not the case with cumulative impacts. Cumulative impacts involve the *combined* effect of the proposed action with other past, present, or reasonably foreseeable actions. Most importantly, the CIA is attempting to reach a decision regarding significance of an impact on a baseline environment that frequently *has already been significantly changed*. Once this difference is appreciated, it comes as little surprise that the assessment of cumulative significance produces so much confusion and frustration. It also explains why the cumulative assessment of GHG emissions will frequently result in the paradox. Unless this paradox is resolved, it will continue to stimulate inefficiencies and needless lawsuits, while actually contributing little value to the decision-making process. What is of importance is developing an approach that leads to meaningful information on which to base a decision without creating worthless paperwork exercises in satisfying paradoxical regulatory requirements.

Complementary interpretation of significance
A solution to the paradox requires that one first understand what is meant by the term "significance." The regulations state that significance must consider both the intensity and context in which the impacts take place.

With respect to intensity, the regulations identify ten factors that should be considered in making a determination regarding the potential significance of an impact (40 CFR 1508.27, Significantly).

Whereas the ten significance factors have been developed to assist agencies in assessing the significance, specific regulatory direction does not exist regarding how they are to be used or interpreted in reaching such determinations. Case law shows that decision makers have, in fact, been given relatively wide latitude in interpreting how these factors are applied in assessing significance. For example, several of the significance factors cited in the regulations state that decision makers should consider the *degree to which* the significance factors may affect the environment (40 CFR 1508.27, Significantly). For instance, one of these significance factors indicates agencies should consider "the degree to which the proposed action affects public health or safety" [10 CFR 1508.27 (b)(2)]. Decision makers, however, have not been given specific direction for interpreting or determining when an impact has affected the environment to such a degree that it constitutes a significant environmental impact; the criteria merely state that agencies should consider the *degree* to which they may affect the environment.

Thus, the interpretation of the ten significance factors has largely been left to the discretion of the agency to *interpret* based on its own internal judgment and expertise. In practice, the individual decision maker is responsible for determining how the CEQ significance factors should be considered. Thus, decision makers must exercise a considerable degree of professional discretion and subjective judgment in making such determinations. An approach is outlined that builds on this practice.

Significant departure principle

The method described in this section is premised on the concept of significance being, to some degree, interpretative. This section introduces a concept referred to as Eccleston's significant departure principle (SDP).[94]

Does the GHG emission significantly affect the baseline?

With respect to the assessment of a cumulative impact, the SDP approach can be viewed in terms of the degree to which a proposed action would change (i.e., depart from) the cumulative impact baseline. In other words, significance can be viewed as the degree to which the impacts of an action would affect or cause a cumulative impact to change or depart from the conditions that would exist if the action were not pursued.

Based on this interpretation, an action could be considered nonsignificant (from the standpoint of cumulative impacts) as long as it does not cause a cumulative impact to depart (relative change) significantly from the condition that would exist if the action were not pursued. Conversely, an impact could be considered significant if it causes one or more cumulative impacts to *significantly change*. Thus, the paradox can be resolved if the significance of a cumulative impact is considered in terms of how much the impact changes (relative change) the cumulative environmental baseline as opposed to the more standard interpretation of assessing the absolute change on an environmental resource.

Such an interpretation is both justified and consistent with NEPA's rule of reason because nothing would be gained, in terms of decision making, by preparing an EIS for an action that would not substantially change the cumulative impact baseline, even if the environmental resource has already been significantly altered. Under the SDP method, a proposal that results in a very small relative change to an environmental resource that has already been significantly impacted may be deemed insignificant. In contrast, a proposed action that would result in a substantial or important relative change to the same environmental resource would be considered significant.

During peer review of this chapter, one criticism of the SDP approach was that a cumulative impact can result from the accumulation of numerous small increments. So how can the SDP approach conclude that a small incremental increase from a proposed action is not cumulatively significant? Indeed, it is true that a cumulative impact can and frequently does result from the accumulation of many small changes. However, as pointed out earlier, such an assessment approach can result in a virtually unworkable paradox. To resolve this paradox, the SDP approach, refocuses the cumulative impact assessment to that of determining if the incremental increase in question is so small, in relative terms, as to add no meaningful change in the baseline condition and thus its contribution can be considered to be nonsignificant. In other words, the focus becomes one of determining if a change is of such a small incremental value as not to constitute a significant change from the baseline conditions. If the incremental value is so small as to have no meaningful change on the baseline condition, it is generally unlikely that preparing an EIS for that proposed action would result in any reasonable alternative or mitigation measure that would be adopted to reduce or avoid such a small impact. Exercised properly, the SDP approach resolves the Paradox and allows analysts to consider cumulative significance in a more reasonable and meaningful manner; it allows decision-makers to focus efforts (e.g., preparation of an EIS and mitigation measures) on proposals where the use of scarce EIS resources (e.g., funding, manpower) can truly make an important contribution to environmental performance.

Example using non-greenhouse emission

Assume that the current 8-hour average baseline concentration of Emission X at a particular site is 190 µg/m³ and that an air quality standard (threshold of significance) calls for a concentration of 160 µg/m³. Because the existing baseline concentration (190 µg/m³) exceeds the regulatory standard (160 µg/m³), it can be viewed as having already sustained a cumulatively significant impact. We now examine four separate cases (Figures 2.3, 2.4, 2.5, and 2.6) involving an increase in this emission. These figures are included for conceptual purposes only, and are not drawn to scale.

Case 1. Assume that an EA is prepared for a project that would involve an increase in concentration of 0.1 µg/m³, resulting in an 8-hour average baseline concentration of 190.1 µg/m³ (Figure 2.3). Because the threshold of significance has already been breached (160 µg/m³), an absolute or standard interpretation of significance suggests that any additional contribution must also be significant (regardless of how trivial an increase); either the emission would need to be mitigated or an EIS would need to be prepared for a project that causes *any* increase in the cumulative impact baseline, regardless of how nominal its effect.

However, the SDP provides a practical alternative interpretation because it allows the decision maker to consider the impact of the proposed project from its incremental (relative) impact rather than from an absolute

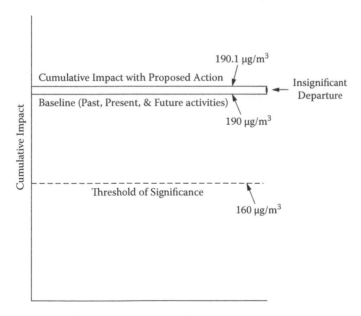

Figure 2.3 Cumulative impact assessment using the significant departure principle.

perspective. That is, it considers whether the *departure or relative change* in the impact is *significant*. If a 0.1-µg/m³ change is considered significant, the entire project is significant, and an EIS must be prepared. Conversely, if a 0.1-µg/m³ change is considered inconsequential and to pose no *significant increase* in the cumulative concentration, then it is deemed nonsignificant and does not require the preparation of an EIS. In this example we assume that the decision maker reviews the finding and concludes that a 0.1% increase is nonsignificant, even though the ambient concentration had already clearly breached the threshold of significance. The reason for this conclusion in simple: The relative contribution is so small as to not add any meaningful change to the worldwide baseline CO_2 level. Thus it can be deemed nonsignificant. This conclusion is in sharp contrast to the more standard (absolute significance) interpretation of significance, which would have triggered the requirement to prepare an EIS. The selection of 0.1% in this example is for illustrative purposes only.

Case 2. Figure 2.4 illustrates a second case where the ambient concentration is 190 µg/m³. But, in this example, the project would increase the ambient concentration by 30 µg/m³, to a total cumulative concentration

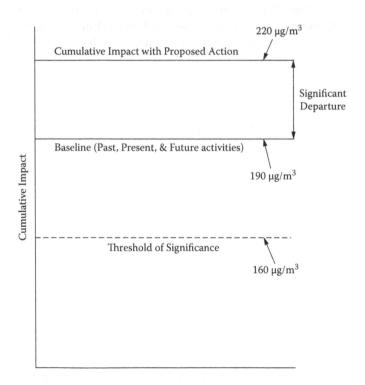

Figure 2.4 Cumulative impact assessment using the significant departure principle.

of 220 µg/m³. The decision maker would be justified in concluding that the project has resulted in a relatively large increase in concentration; this increase represents a significant change in the ambient air quality concentration requiring preparation of an EIS.

Case 3. Under the SDP approach, a cumulative impact does not necessarily have to breach the threshold of significance to be considered significant. Figure 2.5 represents a case where the cumulative concentration baseline is 110 µg/m³, which lies significantly below the threshold of significance (i.e., air quality standard of 160 µg/m³). In this case, however, the proposed project would increase the cumulative impact baseline from 110 to 150 µg/m³, which is still below the regulatory standard (threshold of significance) of 160 µg/m³. Under the more standard interpretation of significance (absolute assessment of significance), it could be, and in fact generally is, concluded that such an increase is nonsignificant and therefore an EIS is not required.

However, the SDP method can lead to the conclusion that a *substantial change* or *departure* in the environmental impact baseline constitutes a significant impact because it would significantly degrade environmental quality, even if it does not actually breach the threshold of significance. Because the proposed action has significantly increased the ambient concentration baseline (even though it has not breached the threshold level of significance), a decision maker could (perhaps should) be justified in concluding that the action constitutes a significant impact requiring the preparation of an EIS. Such a significant change makes it even

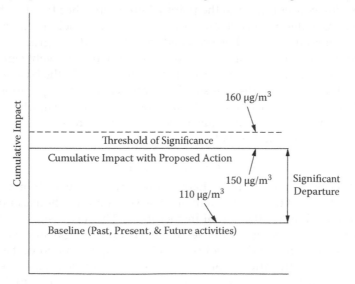

Figure 2.5 Cumulative impact assessment using the significant departure principle.

more likely that future actions will eventually breach the threshold of significance. In this case, preparation of an EIS is warranted because alternatives or mitigation measures might be identified that eliminate or substantially mitigate the substantial increase in the cumulative concentration baseline.

Just as with the more standard interpretation of significance, the SDP assessment of significance is evaluated in terms of the CEQ's ten significance factors and the context in which the impact occurs (40 CFR 1508.27). As an example, significance might be assessed in terms of the degree to which the project would cause a *change* in existing public health and safety [40 CFR 1609.27(b)(2)]. Based on the SDP, an action's effect on public health and safety could be considered nonsignificant as long as the cumulative public health and safety does not *depart significantly* from the conditions that would exist if the action were not pursued. Agency-specific guidance could even be developed to assist decision makers in determining how the ten significance factors, in addition to context, should be interpreted in terms of a relative *change* in the emission or impact.

Case 4. Figure 2.6 represents a special case. In this example, the ambient baseline concentration is 159.9 µg/m3. The proposed action would add a very small incremental increase of .2 µg/m^3. In the other cases described above, an incremental increase of .2 µg/m^3 has generally been assumed to be so small as to not constitute a significant increase (cumulatively significant impact). However, in this case, the increase is just sufficient to increase the baseline concentration to the point where it breaches the threshold of significance—'the preverbal straw that breaks the camel's back.' Because this incremental increase has breached the threshold of significance, it is considered to constitute a significant increase (cumulatively significant impact) and thus requires either mitigation to maintain the baseline concentration level below the threshold of significance or preparation of an EIS. However, if this action does breach the threshold of significance, the contributions from any subsequent Federal actions are subject to the SDP approach described in Cases 1 – 3.

Assessing emission levels versus impacts

During peer review of this chapter, one criticism was that the SDP approach focused too much attention on assessing greenhouse emissions levels rather than on the actual impact itself. This criticism is not without merit. The actual greenhouse emissions and the global greenhouse baseline concentration does not constitute an impact in and of itself. The actual impact is the affect that such emissions produce (e.g., cause a glacier to melt an "x" rate per year, cause precipitation in an area to decrease by "y" amount, or cause a species population reduction of "z" amount). This chapter focuses on GHG emissions because it's conceptually simpler in terms of explaining the SDP concept to the reader. However, provided there is sufficient information available (in most cases there is not sufficiently definitive information to make actual impact projections) the SDP approach would be better applied in terms of the actual assessment of climate change impacts. Assuming there was sufficiently accurate information available, the SDP method could be used to determine if the action would cause a significant increase in an impact such as glacial melting; for instance, assume that as a result of the current global GHG concentration, a major glacier or ice body is currently melting at the rate of 25 meters a year. Now assume that the proposed action would cause this melting rate to increase to 25.00001 meters per year. Under the SDP approach, the cumulative significance question could be framed in terms of whether the additional .0001 meters is significant or nonsignificant. In this case, let us assume that the decision-maker concludes that the cumulative increase in melting (change in melting rate) of .00001 meters is nonsignificant. Now, consider a second case where the proposed action would change the melting rate from 25 meters per year to 25.5 meters. In this case, the decision-maker might well determine that such an increase is cumulatively significant, requiring preparation of an EIS to study alternatives and mitigations measures for reducing the impact.

Comparative summary of the SDP process

Figure 2.7 provides a comparative summary of the SDP approach. This example is for illustrative purposes only, as the purpose of this figure is merely to contrast and summarize some of the succinct concepts described above.

The left side of the figure (Cases 1A, 1B, 1C, and 1D) depict a relatively small incremental increase from a proposed action which has been 'added' to the baseline concentration (baseline contribution from other past, present, and reasonably foreseeable future actions). Similarly, the right side of the figure (Case 2A, 2B, 2C, and 2D) show similar situations, but in these cases, the contribution is a much larger incremental increase in the baseline concentration.

The incremental values in 1B, 1C, and 1D are all equal. Similarly, the incremental values in 2B, 2C, 2D are also all equal. Cases 1A and 2A merely depict a small and a larger baseline; both cases (1A and 2A) assume that the proposed action will not resulted in any incremental increase in either baseline.

Case 1B depicts a much larger baseline than 1A (a situation years into the future when additional contributions have substantially increased the baseline level). The proposed action when added to Case 1B, represents a relatively small increase in the total concentration, and thus a decision-maker would probably be justified in concluding that the incremental contribution is not cumulatively significant and therefore an EIS is not required (note that this case only considers the cumulative GHG emission and not the direct or indirect impacts, or the actual physical impacts that the GHG emission may have on resources such as biota, stream flows, or aesthetics etc.). In case 1C, the incremental increase in GHG concentration is sufficient to breach what has been deemed to be the significance threshold; under the SDP approach this breach is considered to be significant (regardless of how small the incremental increase is), and therefore requires either mitigation or preparation of an EIS. In case 1D the significance threshold has already been breached; in this case, the incremental increase is considered to be of such a small magnitude as not to contribute a cumulatively significant increase to the baseline concentration (even though the significance threshold has been breached), and thus an EIS is not required for this proposed action.

Now we will consider a second analogous situation (right hand side of the figure) where the contributions result in a much larger incremental increase in the baseline concentration. In case 2B, the baseline concentration has been drawn much smaller than in the corresponding case 1B. However, in 2B, the incremental contribution from the proposed action is much larger than that show in 1B; this incremental concentration has substantially increased the total concentration to a value equal to that shown in 1B (which has already been assumed to be nonsignificant). However, in case 2B, the incremental increase is much larger in terms of a relative change; because the relative change in the baseline is so large, the decision-maker would be justified in concluding that this incremental increase is cumulatively significant, requiring mitigation or preparation of an EIS. Now consider Case 2C. As in the case of 1C, the incremental increase in 2C is sufficient to breach the significance threshold, which under the SDP approach is considered to be cumulatively significant. In case 2D, the significance threshold has already been breached; in this case, however, the incremental increase is considered to be of such a large magnitude as to contribute a cumulatively significant increase, thus requiring mitigation or preparation of an EIS.

The sphinx scale: assessing the significance of greenhouse gas emissions

The global nature and complexity of GHG emissions necessitate a slight nuance to the SDP method. This modification is referred to as the sphinx scale, which is described next. The author finalized the concept while touring the Great Sphinx in Egypt. The sphinx solution provides a systematic and defensible method for resolving the GHG paradox. Before describing the sphinx solution, consider the state of the cumulative GHG baseline. Literally hundreds of millions of individual activities (e.g., factories, vehicles, concrete curing, deforestation, cattle ranching) emit CO_2 and other GHGs. Unlike many local airshed problems involving sulfur dioxide, nitrous oxides, and other well-known pollutants, CO_2 emissions tend to remain in the atmosphere for a very long period and contribute to a global rather than a local concentration level; this greatly complicates the complexity of the analysis, including the assessment of significance. This also complicates the complexity of the SDP method. Much of this problem stems from the fact that even the largest GHG-emitting projects generally produce no discernible change in the global baseline concentration. This is because the incremental contribution of almost any imaginable proposal is dwarfed by the effect that hundreds of millions of other combined emitters have on the global concentration level. Moreover, these other transnational sources obey no bounds, yet (with the exception of the global common) a NEPA assessment is not normally required to consider the impacts of actions abroad.

Consider, for example, the GHG emissions from the largest imaginable coal-fired plant. Such emissions are negligible in comparison to the global baseline concentration. In contrast to the examples depicted earlier (Figures 2.3 through 2.5), a human would be unable to detect any discernible change in the baseline on a reasonably drawn graph. Thus, the SDP method would always result in a finding of no significant impact, regardless of the size or complexity of the project. Yet, most seasoned decision makers instinctively understand that it would be unreasonable to conclude that the cumulative GHG emissions from a major coal-fired plant are nonsignificant. As witnessed earlier, this is not necessarily the case with other cumulative emissions, such as local sources of sulfur dioxide, nitrous oxides, or ozone, which tend to be more localized; in fact, they can accumulate to significant and observable increases in the local baseline concentration.

The following method illustrates how a small variation in the SDP approach can resolve this problem. The approach described in Chapter 1 is specifically designed to assess the cumulative significance of GHG emissions (a relatively unique problem); the standard SDP approach described earlier should generally be used in assessing most other cumulative impacts (e.g., land disturbance, loss of grassland or wetlands, noise, non-GHG air pollutants, groundwater usage).

The sphinx solution

In lieu of focusing on the degree to which the baseline CO_2 concentration would change on a graph (no observable change), the decision maker should define criteria for gauging cumulative significance in terms of common activities or projects. The following example illustrates how the sphinx solution provides a gauge or scale for resolving the GHG paradox.

Consider some typical types of activities. At the upper end of the CO_2 emission scale lie projects such as large coal-fired plants. A single coal-fired power plant may produce on the order of 25 billion pounds of CO_2 emissions annually.[95] At the lower end of this spectrum lie very small projects, which may involve relatively trivial emissions such as running a federal vehicle over the course of a year; for instance, a small federal action might involve an inspection team using a passenger vehicle that emits 10,000 pounds of CO_2 annually.[96] A project such as cutting down a certain number of trees (emissions via deforestation) might be chosen as a mid-level significance criterion that lies between a coal-fired plant and vehicular emissions. An *emissions scale* could be constructed, ranging from small to large, depicting average emissions for scores of different types of activities.

Recall that under a more standard interpretation of significance, *any* additional contribution in GHG emission would likewise be considered significant. Now consider how an emissions scale might be used to resolve this paradox. Under the SDP, we will assume that the agency has predetermined (perhaps through an EIS that evaluated GHG emissions of various types of activities and assigned a relative GHG concentration significance level to each action) that, from a relative (change) standpoint, projects that produce CO_2 on the order of vehicular emission are clearly nonsignificant; conversely, projects on the order of coal-fired plants are significant.

The significance of carbon emissions from a proposed action could then be assessed in terms of how much it would contribute relative to these preestablished significance criterion levels. Emitters on the lower end of scale would be deemed to clearly have a nonsignificant CO_2 contribution, whereas those on the upper end would constitute a significant cumulative contribution. Professional judgment would be used to assess the significance of actions that fall toward the middle range of the scale. For instance, a proposed construction project involving concrete might produce 100,000 pounds of CO_2. This is ten times the emission of a typical vehicle but 250,000 times less than a coal-fired plant. Thus, compared to a vehicle, its contribution in terms of the global concentration would be modestly large, but compared to a coal-fired plant it would contribute no appreciable change in terms of the global concentration. The decision maker could defensibly conclude that, with respect to an assessment of GHG emissions, the project has no cumulatively significant CO_2 contribution.

Now consider a proposed small coal-fired plant that would produce 2.5 billion pounds of carbon. It is modestly smaller than a typical coal-fired plant but much greater than a vehicle by orders of magnitude. Relative to a vehicle's emission, it would result in a significant cumulative contribution in CO_2 emissions, even if it would not translate into a visible or discernible change in the baseline concentration on a graph such as that shown in Figures 2.3 through 2.5. The decision maker could again defensibly conclude that, with respect to an assessment of GHG emissions, the project would result in a significant cumulative CO_2 impact.

The sphinx scale

The sphinx scale shown in Figure 2.6 illustrates how the sphinx solution can be used to assist decision makers in accessing the significance of activities that contribute to global GHG concentrations. The vertical emissions scale to the right of the figure compares a spectrum of different types of activities, including their respective average CO_2 emissions in tons per year. They are ranked from the lowest (bottom) to the largest emitters near the top of the emissions scale. The vertical line to the right of the figure, the relative significance scale, provides a gauge of the significance of these emitters, ranging from clearly nonsignificant to very significant near the top of this scale. Consistent with NEPA practice and case law, the significance of each federal action would also be assessed in terms of *context*. [*Note:* This example is for conceptual illustration purposes *only* and does *not* represent actual data or significance levels, which would have to be determined by the agency based on a detailed (NEPA) review of various activities, their average emissions, and how these data are to be interpreted and cross-referenced to provide an indication of their relative significance.]

Application of the sphinx scale

Consider the isolated action involving the use of a vehicle for a year by federal inspection team: A CATX or perhaps an EA would be applied to this effort, and its individual GHG contribution would be gauged in terms of a sphinx scale. If, on the other hand, this same agency had a coordinated proposal for a major project involving hundreds of such vehicles, the GHG emissions would be considered collectively in an EIS, and the impacts would be gauged collectively in terms of the sphinx scale. Another example might involve a coordinated training proposal by the US Army, involving thousands of vehicles over a 5-year period; in this case, the EIS would need to collectively assess the GHG impacts of all these vehicles because the action involves an integrated and comprehensive training proposal; these GHG emissions would be assessed collectively in terms of the sphinx scale.

Adoption of the sphinx scale

One criticism of this approach is that an EIS could not be used to establish a sphinx scale because this could be deemed to constitute "rulemaking." However, there are numerous examples in which federal agencies have prepared EISs and used these analyses to interpret, identify, and establish

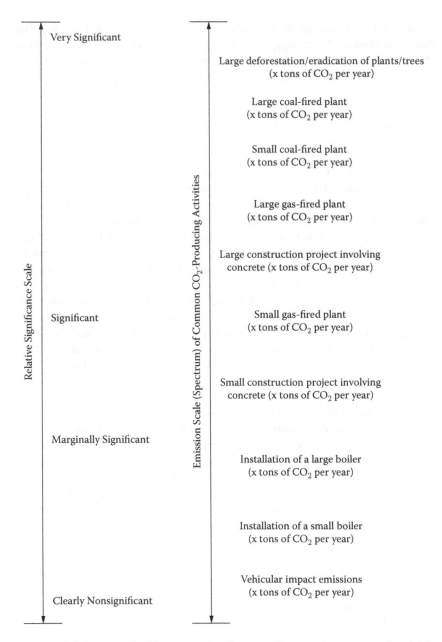

Figure 2.6 Sphinx scale. This example illustrates how various types of activities could be used to provide the decision maker with a relative index for interpreting significance of various greenhouse emission activities. The decision maker would also need to consider the *context* in interpreting the level of significance. This example is for illustration purposes only, and does not represent actual significance levels.

significance indicators and thresholds for later use. Adoption of a sphinx scale could follow a similar process.

Based on its mission, level of expertise, and other factors, each agency might prepare an EIS or even a programmatic EIS to examine GHG significance threshold indicators for use in developing a sphinx scale. Alternatively, the EPA or some other entity might be charged with developing a single sphinx scale that could be used by all federal agencies. Finally, the Council on Environmental Quality or the EPA might consider adopting a sphinx scale pursuant to the Administrative Procedures Act's formal rulemaking process.

Rationale

No discussion of the SDP and its application to the assessment of GHG emissions can be considered complete without addressing the question of defensibility. Specifically, we must consider whether this approach can legitimately be used to resolve the paradox. As confirmed in the following subsections, agencies not only have discretion to adopt methodologies to facilitate the NEPA process, but are encouraged to do so by the CEQ NEPA regulations.

Rule of reason

Judicial decisions have indicated that NEPA is governed by the rule of reason. The term cites a judicial principle intended to[97]

> ... *ensure that common sense and reason are not lost in the rhetoric of regulation, has been invoked to judge the adequacy of an EIS and has been accepted as the appropriate standard.*

The CEQ regulations require agencies to take steps to reduce unnecessary paperwork and delay. As we have witnessed, the aforementioned paradox can lead to a redundant and wasteful process in which an agency is forced to violate the mandate to reduce paperwork and project delays in order to comply with a pedantic requirement that may add little or no benefit to the actual decision-making process.

Little or no useful purpose is served by requiring the preparation of an EIS to evaluate impacts that would not substantially depart from the conditions that would exist if the action were not pursued. Accordingly, it appears unreasonable to require an agency to prepare an EIS for which no useful benefit would be derived. The SDP is consistent with the rule of reason because it provides a systematic and defensible method for addressing the illogical or unreasonable result of the paradox.

Agencies have been granted authority to interpret and determine significance

Significance by its very nature is interpretive. As seen earlier, agencies have been granted a great deal of discretion in interpreting and making determinations regarding significance. The regulations, for example, do not explicitly state how decision makers should view or interpret the degree to which certain significance factors should be considered in determining significance.

When questions have arisen regarding the interpretation of significance, courts have tended to defer in favor of an agency's discretion to interpret significance. For example, the courts have generally given agencies the right to make determinations concerning significance, as long as these determinations are not considered arbitrary and capricious. Daniel Mandelker reports[98]

> *Because a consensus is usually lacking on the state of the art in environmental methodology, the courts have usually accepted the methodology used by an agency in analyzing environmental impacts. They put the burden of proof on plaintiffs to prove the methodology was unacceptable. … These decisions reflect the usual judicial willingness to uphold an agency when the evidence shows that there is only a disagreement among experts.*

Agencies, therefore, appear to be justified in adopting an approach, such as the SDP and its related sphinx solution, as long as it is not used in a manner that is considered "arbitrary and capricious." Accordingly, it appears reasonable that an action can be considered nonsignificant (from the standpoint of cumulative impacts) as long as it does not cause a cumulative impact to depart or significantly degrade conditions that would exist if the action were not pursued.

Agencies granted authority to develop methods for implementing NEPA

The regulations were specifically designed to provide agencies with the flexibility needed to tailor the NEPA process to their own needs. Consistent with this philosophy, the regulations place a minimal number of restrictions on how agencies are expected to implement their NEPA process.

For example, agencies have been given responsibility for developing internal implementing procedures for administering the NEPA process. Specifically, authority has been granted to[99]

> ... *allow each agency flexibility in adapting its imple-*
> *menting procedures* ...

Thus, agencies appear to have been granted discretion to develop a systematic approach for resolving a logical paradox.

Summary

GHG emissions and their effects are becoming the paramount environmental issues of our time. The preparation of NEPA analyses and plans is a key component in assessing and mitigating potential climate change impacts. The more standard interpretation of significance can lead to an unreasonable and absurd conclusion that a FONSI can only be issued if the cumulative impact associated with a proposed action is nonsignificant. Based on this interpretation of significance, many, if not most, activities for which EAs are currently prepared would be technically ineligible for a FONSI, thus requiring the preparation of an EIS, even in cases where the direct and indirect impacts of the proposed activity may be clearly innocuous. Perhaps most troubling, Eccleston's paradox presents an intractable problem that diverts attention and scarce resources from combating truly significant problems to ones that are of little or no importance.

There appear to be only two reasonable solutions for resolving the paradox. One fix involves amending the NEPA (or its regulations) so as to eliminate cumulative GHG analyses or to revise the requirement to ensure that GHG assessments do not lead to the conclusion that every carbon-emitting action requires an EIS in lieu of an EA. However, the author does not believe that such a radical approach is either wise or necessary. The SDP method in combination with the sphinx solution provides a systematic and defensible approach for resolving the paradox.

Distilled to its essence, the difference between the more standard approach and the SDP method is that under the standard approach, significance factors provided in the NEPA regulations (40 CFR 1508.27) are used to assess the significance of a cumulative impact in terms of its *absolute* value or impact; this of course can lead to an impractical paradox. The SDP approach resolves this paradox by interpreting significance factors from a *relative perspective*; under the SDP method, the assessment of significance focuses on determining whether the *change* in the cumulative impact is consequential (i.e., significant). Thus, under the SDP approach, FONSIs can still be issued for a large multitude of relatively low CO_2 emitting activities.

Endnotes

1. Supreme Court decision, *Massachusetts v. EPA*, April 2007.
2. 42 U.S.C. §4332(2)(F).
3. 42 U.S.C. §4321.
4. Synthesis Report at 38 (http://www.ipcc.ch/pdf/assessment-report/ar4/syr/ar4_syr.pdf).
5. 74 Fed. Reg. at 66497–66498.
6. www.ipcc.ch.
7. Intergovernmental Panel on Climate Change, IPCC Fourth Assessment Report: Climate Change 2007, http://www.ipcc.ch/ipccreports/assessments-reports.htm.
8. US Climate Change Science Program, www.climatescience.gov.
9. *City of Los Angeles v. National Highway Traffic Safety Administration*, 912 F.2d 478 (D.C. Cir. 1990).
10. *Border Power Plant Working Group v. Department of Energy*, 260 F. Supp. 2d 997 (S.D. Cal. 2003), *Mid States Coalition for Progress v. Surface Transportation Board*, 345 F.3d 520 (8th Cir. 2003).
11. *Mayo Foundation v. Surface Transportation Board*, 472 F.3d 545 (8th Cir. 2006).
12. *Center for Biological Diversity v. National Highway Traffic Safety Administration*, 508 F.3d 508, 550 (9th Cir. 2007).
13. *Border Power*, 260 F. Supp. 2d at 1028–1029.
14. *Center for Biological Diversity v. Kempthorne*, No. 07-CV-00141 (D. Alaska) transferred from No. 07-CV-0894 (N.D. Cal. filed February 13, 2007.
15. U.S. Dist. LEXIS 53187 (N.D. Cal. 2007).
16. *Border Power Plant Working Group v. Department of Energy*, 260 F. Supp. 2d 997, 1028-1029 (S.D. Cal. 2007).
17. *Center for Biological Diversity v. Kempthorne*, No. C-07-0894-EDL (N.D. Cal.); 2007 U.S. Dist. LEXIS 53187 (N.D. Cal. 2007).
18. *Center for Biological Diversity v. National Highway Traffic Safety Administration* (NHTSA), No. 06-71891 (9th Cir., filed November 15, 2006).
19. *Mid-States Coalition for Progress v. Surface Transportation Board*, 345 F.3d 520, 532, 550 (8th Cir. 2003).
20. *North Slope Borough v. Minerals Management Service*, 2007 U.S. Dist. LEXIS 27487 at *11 (D. Alaska, 2007).
21. *Mosbacher*, 488 F. Supp. 2d at 889; *Montana Environmental Information Center v. Johanns et al.*, No. 07-cv-01311-JR (D.D.C., filed July 23, 2007).
22. No. 3:06CV05755 (N.D. Cal. Sept. 20, 2006).
23. *Ranchers Cattleman Action Legal Fund v. Conner*, No. 07-CV-01023 (D.S.D. filed October 24, 2007).
24. *Center for Biological Diversity v. Kempthorne*, No. 07-CV-00141 (D. Alaska) transferred from No. 07-CV-0894 (N.D. Cal., filed February 13, 2007).
25. *Mid States Coalition for Progress v. Surface Transportation Board*, 345 F.3d 520 (8th Cir. 2003).
26. *Mayo Foundation*, 472 F.3d at 555–556.
27. *Mid States Coalition for Progress v. Surface Transportation Board*, 345 F.3d 520 (8th Cir. 2003), pp. 2–16, *Seattle Audubon Society v. Lyons*.
28. *Association of Public Agency Customers v. Bonneville Power Administration*, 126 F.3d 1158 (9th Cir. 1997).
29. *Mayo Foundation v. Surface Transportation Board* 472 F.3d 545 (8th Cir. 2006).

30. *Montana Environmental Information Center v. Johanns et al.,* No. 07-CV-01311 (D.D.C., filed July 23, 2007).
31. 2007 WL 1577896 (E.D. Cal. May 25, 2007).
32. *Center for Biological Diversity v. Brennan,* 2007 WL 2408901 (N.D. Cal. Aug. 21, 2007).
33. Executive Order 13423, Strengthening Federal Environmental, Energy, and Transportation Management, January 24th, 2007.
34. Executive Order 13514, Federal Leadership in Environmental, Energy, and Economic Performance, 74 FR 52117; October 8, 2009.
35. *Massachusetts v. EPA,* 127 S. Ct. 1438, 1455 (2007).
36. *Deer Creek Shaft and E Seam Methane Drainage Well Project,* Gunnison County, Colorado.
37. CEQ (Council for Environmental Quality), Memorandum for Heads of Federal Departments and Agencies: Draft Guidance for NEPA Mitigation and Monitoring, February 18, 2010.
38. EPA's Mandatory Reporting of Greenhouse Gases Final Rule, 74 FR 56260, October 30, 2009.
39. See 74 FR 56272.
40. 40 CFR 1501.4, 1508.27.
41. *DOT v. Public Citizen,* 541 U.S. 752, 767 (2004).
42. 40 CFR 1500.4(f), (g), 1501.7, 1508.25.
43. 40 CFR 1502.5, 1502.24.
44. 40 CFR 1508.25.
45. *Public Citizen,* 541 U.S. at 768.
46. See http://www.eia.doe.gov/oiaf/1605/FAQ_GenInfoA.htm#.
47. 40 CFR 1500.4(g), 1501.7.
48. 40 CFR 1502.15.
49. 40 CFR 1502.15.
50. *Public Citizen,* 541 U.S. at 767.
51. 40 CFR 1502.21, 1502.22.
52. *Schussman B.,* NEPA Review and impacts on climate change, *CLE International 4th Annual NEPA SuperConference,* March 6 and 7, 2008, San Francisco, March 2008.
53. *Mid-States Coalition for Progress v. Surface Transportation Board,* 345 F.3d 520, 549 (8th Cir. 2003).
54. *Metro. Edison Co. v. People Against Nuclear Energy,* 460 U.S. 766, 774 (1983).
55. *Glass Packaging Institute v. Regan* 737 F.2d at 1091.
56. *Metropolitan Edison v. People vs. Nuclear Energy,* 460 U.S. at 774 n.7.
57. *US Department of Transportation v. Public Citizen,* 541 U.S. 752, 767 (2004) (quoting Keeton et al., *Prosser and Keeton on Law of Torts* 264, 274–275 (5th ed., 1984).
58. *Center for Biological Diversity v. National Highway Traffic Safety Administration,* 508 F.3d at 522-23.
59. *Center for Biological Diversity v. National Highway Traffic Safety Administration,* 508 F.3d at 547, fn. 68.
60. *Friends of the Earth, Inc. v. Mosbacher,* 488 F. Supp. 2d 889 (N.D. Cal. 2007).
61. *No GWEN Alliance of Lane County, Inc. v. Aldridge,* 841 F.2d 946, 952 (9th Cir. 1988) (citing *Trout Unlimited v. Morton,* 509 F.2d 1276, 1283 (9th Cir. 1974)).
62. *Center for Biological Diversity v. National Highway Traffic Safety Administration (9th Cir., Nov 15, 2007).*

63. United States Department of Energy, Recommendations for the Preparation of Environmental Assessments and Environmental Impact Statements, Office of NEPA Oversight, May 1993.
64. *New York v. Kleppe*, 429 U.S. 1307, 1311 (1976), citing *Natural Resources Defense Council v. Calloway*, 524 F.2d 79, 88 (2d Cir. 1975)
65. 51 Fed. Reg. 15,618, 15,621, Apr. 25, 1986.
66. Wishnie, L., NEPA for a new century: climate change and the reform of the National Environmental Policy Act, *N.Y.U. Environmental Law Journal*, 16(3), 646, June 26, 2008, http://www1.law.nyu.edu/journals/envtllaw/issues/vol16/Wishnie.pdf (accessed June 17, 2009).
67. *Metropolitan Edison Co. et al. v. People Against Nuclear Energy*, 460 U.S. 766, 774, (1983).
68. *Glass Packaging Institute v. Regan*, 737 F.2d at 1091.
69. *Metropolitan Edison v. People vs. Nuclear Energy*, 460 U.S. at 774, fn 7.
70. *U.S. Department of Transportation v. Public Citizen*, 541 U.S. 752, 767 (2004) (quoting W. Keeton et al., *Prosser and Keeton on Law of Torts* 264, 274–275 (5th ed. 1984)).
71. *No GWEN Alliance of Lane County, Inc. v. Aldridge*, 841 F.2d 946, 952 (9th Cir. 1988) (citing *Trout Unlimited v. Morton*, 509 F.2d 1276, 1283 (9th Cir. 1974)).
72. Moore, C. G., III, Allen, L. G., and Forman, M. R., Indirect impacts and climate change: assessing NEPA's reach, *Natural Resources & Environment*, 23(4), Spring 2009.
73. *Mid-States Coalition for Progress v. Surface Transportation Board*, 345 F.3d 520 (8th Cir. 2003).
74. *Sierra Club v. Marsh*, 976 F.2d 763, 767 (1st Cir. 1992).
75. *Department of Transportation v. Public Citizen*, 541 U.S. 752 (2004).
76. DOE, NEPA Lessons Learned, NEPA and Climate Change: 'Don't Do Nothing', p. 19, June 2009.
77. FEIS for the Sierra Nevada Forest Plan Amendment (January 2004).
78. *National Audubon Society v. Kempthorne*, Case No. 1:05-CV-00008-JKS, Mem. Decision, Dkt. # 107, at 17 (D. Alaska, 2006).
79. FEIS for the Imperial-Mexicali 230-kV Transmission Lines, Vol. 1 at 4-32 to 4-34, 4-57 to 4-59, *www.http://web.ead.anl.gov/bajatermoeis/* ; FEIS for the Imperial-Mexicali 230-kV Transmission Lines, Vol. 1 at 4-32 to 4-34, 4-57 to 4-59, *www.http://web.ead.anl.gov/bajatermoeis/*.
80. DOE, DOE/EIS-0394, 2007.
81. DOE, DOE/EIS-0357, 2007.
82. See *Friends of the Earth v. Mosbacher*, Civ. No. C02-4106, JSW, Plaintiffs' Cross Motion for Summary Judgment; Opposition to Motion for Summary Judgment (filed N.D. Cal., February 11, 2005).
83. See *Mayo Foundation v. Surface Transportation Board*, 472 F.3d 545, 555-56 (8th Cir. 2006); the Eighth Circuit held that Surface Transportation Board's EIS adequately analyzed air impacts even while the EIS stated that impacts of air contaminates, such as greenhouse gases, were too speculative to evaluate. This case preceded the Supreme Court's *Massachusetts v. EPA* decision.
84. 40 CFR §1508.8(a) and (b), §1508.25(c), §1508.27(b)(7).
85. 40 CFR §1508.7.
86. CEQ (Council for Environmental Quality), Considering Cumulative Effects under the National Environmental Policy Act, 1997.

87. 488 F. Supp. 2d 889 (N.D. Cal. 2007).
88. Case No. 07-cv-01311-JR, complaint filed July 23, 2007.
89. *Kleppe v. Sierra Club*, 427 U.S. 390, 412 (1976).
90. *Oregon Natural Resources Council v. Marsh*, 52 F.3d 1485 (9th Cir. 1995) (did not consider effects of all dams in the area, which violated the agency's cumulative impact mandate); *Cascadia Wildlands Project v. U.S. Forest Service*, 386 F. Supp. 2d 1149, 1167–68 (D. Or. 2005); as the distance between projects and lack of a direct relationship between them, a cumulative impacts analysis was unnecessary; *California v. U.S. Department of Transp.*, 260 F. Supp. 2d 969, 974–78 (N.D. Cal. 2003) (lack of a cumulative impact analysis was violated by NEPA because other current and reasonably foreseeable projects in the area of the project were not assessed).
91. *76 City of Oxford v. Federal Aviation Administration*, 428 F.3d 1346, 1353 (11th Cir. 2005); *Blue Mountains Biodiversity Project v. Blackwood*, 161 F.3d 1208, 1214 (9th Cir. 1998); *City of Shoreacres v. Waterworth*, 420 F.3d 440, 453 (5th Cir. 2005).
92. Wishnie, L. G. 2008 NEPA for a new century: Climate change and the reform of the National Environmental Policy Act. NYU *Environmental Law Journal* 6(3), p. 646.
93. Swartz, L., *National Association of Environmental Professionals 35th Annual Conference*, session titled Recent NEPA Cases, Session #5, Atlanta, GA. April 27–30, 2010.
94. Eccleston, C. H. 2006. Applying the significant departure principle in resolving the cumulative impact paradox: Assessing significance in areas that have sustained cumulatively significant impacts. *Journal of Environmental Practice* 8(4): 241–250.
95. Levin, I. 2007. Dirty Kilowatts: America's Most Polluting Power Plants. Environmental Integrity Project, Washington, D.C. 56 pp.
96. U. S. Environmental Protection Agency. 2000, April. Emission Facts: Average Annual Emissions and Fuel Consumption for Passenger Cars and Light Trucks. EPA 420-F-00-013.
97. Freeman, L. R., March, F., and Spensley, J. W., *NEPA Compliance Manual*, Rockville, MD: Government Institutes, Inc., 1992.
98. Mandelker, D. R., *NEPA Law and Litigation (2nd edition)*, Chapter 10, St. Paul, MN: Clark, Boardman, and Callaghan, 1993.
99. 40 CFR 1507.1.

chapter three

Preparing risk assessments and accident analyses

Ignore the environment. It'll go away.

—(Seen on a bumper sticker)

Most people understand that there is a small but finite risk of being killed by a falling airplane. While most people do not view this as a significant risk, they are surprised to learn that the risk of being killed by a crashing aircraft is four times the threshold risk (one chance in a million) that the US Environmental Protection Agency (EPA) uses in evaluating standards for regulating toxins.[1] Even more surprising is how few people realize the risk of a fatality due to a large comet or asteroid impact.

Many geologists now believe that an asteroid impact led to the demise of the dinosaurs.[2] The Earth stands approximately a one-in-a-million chance per year of being struck by a massive object on the order of 2 kilometers in diameter. Such an event would cause extensive human mortality across the planet on a cataclysmic scale. The risk to an individual is on the order of about 1 in a million per year, or 1 in 20,000 per 50-year lifetime.[3, 4] To put this in a risk assessment perspective, this is 50 times larger than typical risks considered by the EPA, and over 250 times larger than the conservative risk standards used by the EPA. When an environmental risk this large comes to the attention of the EPA, it is certain to receive regulatory attention and, very likely, press coverage.[5] Yet the public displays no signs of being particularly alarmed at the much larger odds of being struck by an airplane or asteroid.

This is particularly interesting given the fact that such an astronomical disaster might be avoidable, which raises questions about risk and resource priorities. An astronomical survey system could be developed to provide an early warning system for detecting wayward bodies. Many technologies might be developed for "nudging" a newly discovered asteroid into a different trajectory that would avoid the Earth. It would be expensive, to be sure, yet no popular resounding call has been made to develop such a system. This seems odd indeed, given that the American public (and many other developed nations) seems willing to spend almost unlimited sums on disposal of nuclear waste, which carries an infinitesimally smaller risk than a colliding asteroid.

In *The Gift of Fear*, Gavin de Becker argues that "True fear is a gift. It is a survival signal that sounds only in the presence of danger. Yet unwarranted fear has assumed a power over us that it holds over no other creature on Earth. It need not be this way." A risk assessment provides a rigorous method for collectively measuring and sharing this "true fear." Such assessments can be instrumental in recognizing and assessing irrational biases and other factors that influence decision making and result in misallocated funds. In so doing, it can significantly reduce the potential for disasters caused by naive intuition disguised as rational thinking.

This chapter describes the process and problems commonly encountered in preparing risk assessments and accident analyses. We begin this chapter with a discussion of risk assessment.

Definitions of risk

Numerous definitions of risk are in use. The definition varies depending on the user. Covello and Merkhofer define risk as "the possibility of an adverse outcome, and uncertainty over the occurrence, timing, or magnitude of that adverse outcome."[6] Many scientists define risk as the nature of the harm that may occur, the probability that it will occur, and the number of people that will be affected.[7]

The public, on the other hand, may be more concerned with programmatic, qualitative attributes, such as the origin of risk (natural versus man-induced), whether a risk is imposed or can be voluntarily assumed, the equitable distribution of risk over a population, and the power of individuals to control the risk.[8] The inconsistent and ambiguous use of the word has been the source of significant criticism and confusion.[9]

In its more formal and quantitative usage, risk is proportional to both the expected result of an event and to the probability of the event. In other words, "Risk is a combination of the likelihood of an occurrence of a hazardous event or exposure(s) and the severity of injury or ill health that can be caused by the event or exposure(s)."[10] Mathematically, this concept is simply defined as

Risk = (Probability of an accident) × (Losses per accident)

Or more generally,

Risk = (Probability of event occurring) × (Impact of event occurring)

However, there are more sophisticated definitions also in use. Methods have also been developed to compute the cost of potential loss of human life. One method has involved assessing the cost and extent to which people insure themselves against death (life insurance).[11]

Terms such as "likely," "unlikely," "common," or "rare" are qualitative probabilities whose interpretation varies according to the context in which they are used.

Risk assessment

"Risk assessment" can be defined as "[t]he process of identification, quantification, evaluation, acceptance, aversion, and management of risk." The concept of risk can generally be divided into three tiers:

- In the upper tier, risks are regarded as completely unacceptable and must be reduced even at very high cost.
- The intermediate tier is one in which risk reduction decisions are made by trading off associated costs and benefits.
- In the lower tier, the public readily accepts risks because benefits associated with the risk are believed to outweigh the disadvantages.

The term "risk management" applies to the managerial response based on the resolution of various policy issues such as acceptable risk. Decisions regarding risk management are generally reached by considering a risk assessment in terms of the context of political and socioeconomic constraints. Such decisions are frequently very controversial due to variations in the public's perception of risk.

The concept of *risk acceptance* implies that an entity is willing to accept some risks to achieve some gain or benefit, assuming the risk cannot be practically or easily avoided. The acceptance level is a reference level against which a risk is compared. If the risk level is below the acceptance level, the risk is deemed acceptable. Conversely, if it is deemed unacceptable and avoidable, steps may be either taken to control the risk or forego the activity.

The public's perception and acceptance of risks varies with the nature of the risks and depend upon many underlying as well as psychological factors. Common factors include, but are not limited to, whether the risk

- Is viewed as a "dreaded" hazard or a common hazard
- Is encountered occupationally or non-occupationally
- Results in immediate or delayed effects
- Affects average or especially sensitive people

Risk aversion involves controls that are taken to

- Avoid or reduce the risk
- Reduce the vulnerability of exposed persons

- Regulate or modify the activities to reduce the magnitude or frequency of adverse effects
- Implement mitigation and recovery procedures
- Implement loss-reimbursement schemes and other factors

The process for assessing risk

Risk assessment provides one approach for evaluating impacts of proposals and reaching decisions based on calculated risk. Peter Montague provides a good introduction to this subject in his essay entitled *Making Good Decisions*.[12] Risk assessment has been described by the National Academy of Sciences (NAS) as a four-step process outlined in Table 3.1.[13] To more clearly illustrate this concept, the following discussion focuses on describing risk assessment for a chemical or radiological exposure.

Table 3.1 The National Academy of Sciences Four-Step Process for Assessing Risk

1. **Hazard identification.** The initial step involves estimating risks from specific effects (e.g., chemical or radiological damage from a short-term acute dose, or long-term chronic exposure). For example, a toxic endpoint for a chemical or radiological exposure might include cancer, damage to the central nervous system, damage to organs, developmental disorders, or damage to the immune system or genes.

   Because test organisms (whether rats or people) react differently at various stages of development (particularly in the womb), dozens of endpoints must be considered. In actual practice, many endpoints are simply ignored.

2. **Dose-response assessment.** The term generally refers to the science of determining the degree of damage or impact that will occur from a given chemical or radiation dose. Typically, an increased dose leads to greater effects. Assessing a radiation or chemical dose response typically requires extrapolating from data about laboratory animals, which have been given high doses, to effects in humans who typically receive much lower doses. While such extrapolations open the door to many errors, they often provide the only reasonable approach for determining potential impacts.

3. **Exposure assessment.** The term refers to the science of determining how much of a chemical is absorbed from all sources. Consider lead. Exposures can occur through air, ingestion of water and food, and skin absorption. In practice, however, many potential exposure pathways may be ignored.

4. **Risk characterization.** The final step uses the information gleaned from the first three steps, with information about the characteristics of the affected population (age, health status, race, weight). In practice, characteristics of the receptor population are often ignored and averages are instead used. Taken together, this information attempts to estimate the overall hazard or risk.

Some critics have charged that, as currently practiced, the assessment of risk is often overly simplified and unrealistic; some of these shortcomings are briefly noted in Table 3.1.

The resulting risk is generally expressed as a probability of given hazard on a specified population over a given time. For example, the risk might be expressed in terms of the following statement:

> Over an average lifetime of 70 years, the population within 100 miles of the proposed manufacturing facility can be expected to endure one additional cancer for every 1,000,000 people (above the normal risk of cancer), as a result of chronic exposure to breathing Agent Q.

Flawed or misleading risk-based assessments

Because people are typically exposed to multiple chemicals simultaneously, a risk assessment never fully describes reality. Yet the results often tend to be treated as if they describe the real world. Thus, many critics charge that the NAS risk assessment model outlined in Table 3.1 is overly simplified.

Critics also charge that there are no agreed-upon methods for assessing damage to various organs or genes.[14] Furthermore, the art of risk assessment generally lacks techniques for evaluating exposure to multiple chemicals simultaneously, and therefore may overlook synergistic effects. Risk-based assessments are therefore often misleading, either overestimating or underestimating the safe level of exposure to a toxin. The results can lead to false assurances of safety while allowing damage to occur. Conversely, they can lead to unfounded fears and misallocation of money and resources.

Critics have responded that "risk is unacceptable if it is avoidable." Yet, most risk assessments typically do not reach a conclusion that a risk is avoidable because the risk assessment process generally fails to ask whether a particular risk can be avoided.

Vulnerability assessment

From the perspective of disaster management, "vulnerability" means assessing the threats from potential hazards to the population and to the infrastructure. A vulnerability assessment involves the process of identifying, quantifying, and prioritizing the vulnerabilities in a system. Examples include nuclear power plants, energy supply systems, communication systems, transportation systems, and water supply systems. Such assessments can be performed in terms of political, social, economic, and environmental disciplines.

In general, a vulnerability analysis serves to categorize key assets and drive the risk management process. The US Department of Energy reports that[15]

> ...risk analysis is principally concerned with investigating the risks surrounding physical plant (or some other object), its design and operations. Such analyses tend to focus on causes and the direct consequences for the studied object. Vulnerability analyses, on the other hand, focus both on consequences for the object itself and on primary and secondary consequences for the surrounding environment. It also concerns itself with the possibilities of reducing such consequences and of improving the capacity to manage future incidents.

A vulnerability assessment has a number of traits in common with risk assessment. These assessments are typically reformed using the following steps:

1. Identify and catalog assets and capabilities (resources)
2. Assign quantifiable value or importance (or at least rank order) of these resources
3. Identify the vulnerability threats to each resource or discipline
4. Mitigate the most serious vulnerabilities

A combined NEPA and risk assessment process

In the past, decisions have been made (sometimes with little or no public input) to incinerate highly toxic waste that could affect members of a nearby community. Such decisions are sometimes made based almost solely on the results of a risk assessment. For example, a consultant may be retained to prepare a risk assessment that demonstrates that a proposed facility will only harm a miniscule number of residents. Because the potential harm is so small, the facility is deemed to be acceptable.

Now consider a similar process, but this time, one performed pursuant to the US National Environmental Policy Act (NEPA) or an equivalent international environmental impact assessment (EIA) process. NEPA, and most EIA processes require an open, objective, and democratic process for reaching decisions; it blends public participation and consideration of all reasonable alternatives (including that of taking no action, i.e., avoiding a proposal or given risk) into the standard risk assessment process.

Under NEPA, a problem can be more openly framed and studied by asking which *alternative* would satisfy the underlying need for taking action, while also minimizing adverse impacts? Under NEPA, the process can be more clearly focused: what alternatives and mitigation measures should be studied for satisfying the underlying need (which includes minimizing adverse impacts)? Thus, NEPA offers a more democratic and open process that can objectively compare a wide range of potential alternatives.

Suppose, for example, that an issue is raised regarding the disposition of a particularly toxic waste. Under NEPA, the scope of potential alternatives considered is often much broader as compared to a stand-alone risk assessment process. For instance, potential NEPA alternatives might include banning the use of the toxin altogether; instituting a waste-recycling or minimization program that minimizes a toxin; changing the manufacturing process to avoid generating this waste; or even shipping the waste to a facility where it can be handled more safely.

Dealing with missing or incomplete information in an EIS

An agency may find that information needed for the evaluation of impacts, particularly those related to a risk assessment or accident analysis in a NEPA environmental impact statement (EIS) cannot be obtained because the overall costs of doing so are exorbitant or the means to obtain such information are unknown. The NEPA implementing regulations (40 Code of Federal Regulations [CFR] §1502.22) specify how an agency is to proceed under such circumstances. This provision states that when an agency is evaluating reasonably foreseeable significant adverse effects on the human environment and there is incomplete or unavailable information, it must make clear that such information is lacking. If relevant incomplete information is essential in making a reasoned choice between alternatives and the overall costs of obtaining it are not exorbitant, the agency is required to include the information in its analysis. If such information cannot be obtained because the overall cost of obtaining it is exorbitant or the means to obtain it are not known, then the agency is required to publicly include the following steps within its EIS:

1. A statement that such information is incomplete or unavailable
2. A statement of the relevance of the incomplete or unavailable information to evaluating the reasonably foreseeable significant adverse impacts on the human environment
3. A summary of existing credible scientific evidence that is relevant to evaluating the reasonably foreseeable significant impacts on the human environment
4. The agency's evaluation of such impacts based on theoretical approaches or research methods generally accepted in the scientific community

For the purposes of the US NEPA implementing regulations (40 CFR §1502.22), the term "reasonably foreseeable" includes impacts that have catastrophic consequences, even if their probability of occurrence is low, provided that the analysis of impacts is

1. Supported by credible scientific evidence
2. Is not based on pure conjecture
3. Is consistent with the rule of reason

As discussed below, NEPA's provisions can be particularly important in developing an analysis of accident-related risks. An EIS or other environmental study previously prepared by a federal agency can be used to assist in complying with the requirements of NEPA. In fact, the NEPA regulations encourage use of other related studies.[16] An agency, however, must independently evaluate any document (including an EIS, environmental assessment [EA], or other environmental report) prepared by others that the agency intends to rely on.[17] If such analyses satisfy an agency's obligation to study the potential effects of its own proposed action, then the agency has no obligation to prepare its own study. However, an agency may not substitute compliance with standards or regulations administered by another agency in lieu of a required NEPA analysis.

Ecological risk

The assessment of ecological risk in terms of chemical impacts is frequently more difficult to perform when compared to investigating the impacts on humans. Human risk assessment deals with only one species—humans. In contrast, ecological risk assessments usually consider a wide range of organisms (very small to large); aquatic animals, as well as those that fly or live on the ground; and species with short-life spans and those with long ones. Consequently, there is much greater variability (less standardization) in terms of uncertainties. One source of uncertainty is that of determining the degree of exposure to a species as it migrates across a habitat. There are also uncertainties concerning how accurately an artificial laboratory situation represents the real world. Evaluating the amount of toxin required to produce an effect is a priority for future research. A commonly accepted, comprehensive chemical assessment of risks to the non-human biota is unlikely to be available for some time.

Guidance memorandum by Presidential Commission on Risk

The US Congressional Office of Management and Budget (OMB) and the Office of Science and Technology Policy have recently issued a joint memorandum entitled *Updated Principles for Risk Analysis*. This guidance elaborates on generally accepted principles for risk analysis.[18] The memorandum notes that risk assessment "is not a monolithic process or a single method" and that "... risk assessments share some common principles, but their application varies widely among domains." The memorandum references a 1997 report by the US Presidential Commission on Risk, which stated that the degree of effort expended in performing a risk assessment should be

> *...commensurate with the problem's importance, expected health or environmental impact, expected economic or social impact, urgency, and level of controversy, as well as with the expected impact and cost of protective measures.*

Some of the memorandum's key recommendations include the following:

- *Concise Executive Summary:* A concise summary can improve the clarity of a risk-based analysis and can help ensure that readers interpret it accurately. This summary could disclose the
  - Objectives and scope of the risk assessment
  - Key findings of the analysis
  - Principal scientific limitations and uncertainties
- *Sliding-Scale Approach:* While this memorandum does not use the term "sliding scale," it does reaffirm the principle that the scope of a risk analysis should correspond to the nature and significance of the decision to be made. The US Department of Energy states that a sliding-scale approach:

> *Recognizes that agency proposals can be characterized as falling somewhere on a continuum with respect to environmental impacts. This approach implements CEQ's instruction that in EISs agencies "focus on significant environmental issues and alternatives (40 CFR 1502.1) and discuss impacts 'in proportion to their significance' (40 CFR 1502.2[b]). The reader should note that under CEQ's regulations and judicial rulings, a factor in determining significance involves the degree to which environmental effects are likely to be controversial with respect to technical issues.*

- *Use Best Available Methodologies:* The memorandum states that "Agencies should employ the best reasonably obtainable scientific information to assess risks to health, safety, and the environment...." It goes on to note that "... analyses should be based upon the best available scientific methodologies...." Furthermore, it notes that, "... characterizations of risks ... should be both qualitative and quantitative, consistent with available data."

The memorandum emphasizes the importance of acknowledging and consistently communicating the uncertainties of risk assessments. To this end, "Judgments used in developing a risk assessment, such as assumptions, defaults and uncertainties, should be stated explicitly. The rationale for these judgments and their influence on the risk assessment should be articulated."

The memorandum cautions against presenting single projections of risk because they can be misleading and may provide a false sense of precision. Instead, it suggests that a range of plausible risk estimates or scenarios be presented, and that, when possible, quantitative uncertainty analysis, sensitivity analysis, and a discussion of model uncertainty be included.

Finally, the memorandum notes the importance of addressing the range of scientific and technical opinions and opposing views. Results based on different effects or different studies should be presented. It notes that when relying on data from one study over others, a clear rationale should be provided to justify why this was done. A high degree of transparency with respect to data, assumptions, and methods increases the credibility of the risk analysis, and will allow interested individuals to better understand the technical basis of the analysis.

Uncertainty versus risk

Economist Frank Knight established the important distinction between *risk* and *uncertainty*:[19]

> *Uncertainty must be taken in a sense radically distinct from the familiar notion of risk, from which it has never been properly separated.... The essential fact is that 'risk' means in some cases a quantity susceptible of measurement, while at other times it is something distinctly not of this character; and there are far-reaching and crucial differences in the bearings of the phenomena depending on which of the two is really present and operating... It will appear that a measurable uncertainty, or 'risk' proper, as we shall use the term, is so far different from an unmeasurable one that it is not in effect an uncertainty at all.*

This concept has also found its way into some risk assessments. Hubbard defines uncertainty and risk differently:[20] *Uncertainty* simply means the lack of certainty; a state of having limited knowledge such that one cannot exactly describe an existing state or future outcome. He describes uncertainty and risk as follows:

- *Measurement of uncertainty:* Defines a set of possible states or outcomes, with probabilities assigned to each possible state or outcome
- *Risk:* A state of uncertainty where some possible outcomes cause an undesired effect or significant loss
- *Measurement of risk:* A set of measured uncertainties where some possible outcomes are losses and the magnitudes of those losses are defined

Risk in decision making

"Perception of risk" refers to how individuals or society as a whole anticipate the outcomes of choices made by ourselves or others. The perception of risk has much to do with how it is viewed and interpreted. Many factors influence how people perceive and respond to risk, including an individual's values, beliefs, and attitudes, as well as wider social or cultural values.

According to Covello, factors that influence the perception of risk include features such as perception of risk, amount of control over the risk, benefits accrued from accepting risk, and most importantly trust.[21]

Factors affecting perception of risk and risk-based decision making

"Loss aversion" refers to the tendency of most people to strongly prefer avoiding losses over acquiring gains. Some studies suggest that losses are psychologically twice as powerful as gains.

It is not the reality of a loss that is important, but the perception. Nations have gone to war because of loss aversion. As one textbook states, "Once we have committed a lot of time or energy to a cause, it is nearly impossible to convince us that it is unworthy."[22] The real question ultimately becomes "How bad do your losses have to be before you change course?" *Cognitive dissonance* provides a possible explanation of this effect (described in more detail later). For more information, the reader is referred to the author's companion book entitled *Global Environmental Policy*.[23]

Risk aversion and irrational decision making

Regret can play a significant role in decision making, distinct from *risk aversion* (preferring the status quo to a situation where the risk could make one worse off). Much economic thought has revolved around the rationality of decision making. The relatively new science of behavioral economics has challenged this rational choice theory. Rational choice theory dictated that people act rationally, making choices that efficiently maximize profit. The theory held that when faced with an economic decision, we consider the value of an outcome and make a rational decision about the best or most efficient course to take in maximizing utility or profit. But more recent research reveals that many of our economic choices are driven not by rational calculations, but by deep and unconscious emotions. Consider the following experiment from the field of economics.

1. In the first scenario, I give you $100 and a choice between (A) a guaranteed gain of $50, and (B) a coin flip, in which heads gets you another $100 or tails loses you $100. Do you select (A) or (B)? Behavioral economists have discovered that most people choose (A).
2. Now imagine a second scenario in which you are given $200 and a choice between (A) a guaranteed loss of $50, and (B) a coin flip, in which you lose $100 with a head or nothing with a tail. Do you select A or B?

The final outcome for (A) in each scenario is the same (winding up with $150); this is also the final outcome for (B) (an even chance of winding up with $100 or $200, that is, a 50% chance of winning $150). So rational choice theory predicts that whether you are in the first scenario or the second, you should make the same decision.

Yet, studies indicate that most people choose (A) in the first scenario (a sure gain of $50) and (B) in the second scenario (avoiding a sure loss of $50). This is true although there is no rational difference between having $100 with a sure gain of $50, or having $200 with a sure loss of $50. However, there is an *emotional* difference. That tendency is referred to as *risk aversion*, and hundreds of experiments have demonstrated that most people have an aversion to risk.

People generally tend to reject the prospect of a 50% probability of gaining or losing money unless the amount to be gained is at least double that to be lost. This leads to the conclusion that losses hurt twice as much as gains feel good; only when the potential payoff is more than double the potential loss will most people make the investment gamble. Our emotions are actually driving our risk-taking decisions. There is good reason to believe that the principle of risk aversion applies to any number of other disciplines, including that of risk decision making.

Judgment by heuristic
David Kahneman became the first psychologist to be awarded the Nobel Prize in economics for his work on applying cognitive behavioral theories to decision making in economics. He studies the question of decision

making through "judgment by heuristic." The word "heuristic" means "rules of thumb" (i.e., rules that assist us in the process of learning and making decisions). Kahneman was interested in finding out to what extent our decision making was influenced by our perceptions, emotions, and prior knowledge.

Kahneman believes that factors such as the influence of our emotions, intensity of our fears, possible outcomes and opportunity costs all cloud our judgment and thus our decisions may be irrational. Government decisions on allocation of resources and so on could all be affected by such heuristics as well. What may be a rational and logical choice, is often not the principle factor considered in reaching a decision.

We draw inferences and judgments about factors that can affect our decisions and many of these judgments are completely irrational but we will make them all the same. Irrational responses such as this may explain how the public views and accepts or rejects certain types of risk.

Cognitive dissonance

As alluded to earlier, *cognitive dissonance* refers to an uncomfortable feeling caused by holding two contradictory ideas simultaneously. These ideas or cognitions may include attitudes and beliefs, as well as awareness of one's behavior. *Dissonance* occurs when a person perceives a logical inconsistency among cognitions, such as one idea implying the opposite of another. For example, a belief in animal rights would be inconsistent with eating meat. But some people do so nevertheless. The contradiction leads to dissonance, which can cause guilt, shame, embarrassment, or other negative emotions.

A particularly powerful cause of dissonance is when an idea conflicts with a fundamental belief such as, "I made the right decision." The theory holds that people have a motivational desire to either reduce dissonance by changing their beliefs, attitudes, or behaviors, or to justify and rationalize these traits.[24] This frequently leads to denial of the evidence, or *confirmation bias*, and other similar defense mechanisms. Confirmation bias is the tendency to interpret information in a way that confirms one's preconceptions, and avoids information and interpretations that contradict prior beliefs. Such factors may have a profound influence on risk-based decision making.

Risk homeostasis hypothesis

The perception of risk may play a pivotal role in the interpretation of uncertain information, and risk acceptance or avoidance may be highly influenced by the degree of uncertainty. Understanding risk assessment or avoidance is part of the more general question of how cognitive behavior—both individual and social—may respond to uncertainty in a changing world.

The *risk homeostasis* hypothesis proposed by Gerald Wilde asserts that everyone has a fixed level of acceptable risk. As the level of risk in one part of an individual's life increases, it will tend to trigger a corresponding decrease of risk elsewhere, bringing overall risk back into equilibrium.[25]

One often-cited study involves taxi drivers in Germany. Half the fleet of taxicabs were equipped with anti-lock brakes. The other half drove cars with conventional brake systems. Yet surprisingly, the crash rate was the same for both groups. Wilde asserts that this was due to ABS-equipped drivers assuming that the breaking system would protect them, and consequently taking more risks; non-ABS drivers, in contrast, drove more carefully because they had no ABS to protect them. Wilde argues that the significant increase in car safety features has had little effect on the rate overall crash rate, as safety programs merely shift rather than reduce risk.

Another study found that drivers tend to behave less carefully around bicyclists wearing helmets than around unhelmeted bicycle riders. If true, this begs the following question. If a decision maker chooses a less risky alternative, does that imply that other, more risky decisions will be made, such that the overall risk remains essentially the same? Does it also imply that the public's acceptance of the future risk could also be equally skewed?

No action

Some studies indicate that people tend to have a preference for maintaining the status quo (i.e., no-action alternative).[26, 27] They may even prefer a riskier situation to a less risky situation, if the former maintains the status quo.[28] That is, people tend to be more averse to a risk incurred by taking an action than a similar risk incurred by taking no action. For example, a University of Pennsylvania study found that nonvaccinators (parents who chose not to vaccinate their children) were more likely to accept deaths caused by a disease (i.e., omitting vaccination) than deaths caused by vaccination (an act of commission).[29] This begs another troubling question. As a society, could we be endangering ourselves through a tendency to take no action, as compared to taking an action that might result in an overall reduction in risk?

Optimism bias and planning fallacy

Optimism bias is a demonstrated and systematic tendency for people to be overly optimistic about the outcome of a planned action, particularly with respect to estimates of benefits, costs, and schedule. It must be accounted for explicitly in proposed actions if these are to be realistic. Many studies have shown that optimism bias can cause people to take imprudent or even unacceptable risks.[30] This effect, however, is not universal; for instance, some people tend to overestimate the risks of low-frequency events, particularly negative ones.

Flyvbjerg argues that what appears to be optimism bias may, on closer examination, be strategic misrepresentation.[31] He believes that planners frequently underestimate costs and overestimate benefits deliberately in order to get their projects approved, especially when projects are large and organizational and political pressures are high. Kahneman and Lovallo disagree, maintaining that optimism bias is the main problem.[32]

Rational and irrational judgments concerning risk

Described in the following sections are factors that can influence one's assessment, interpretation, and judgment of risk. Much of the remainder of this chapter deals with rational and irrational factors that can influence or completely skew judgments concerning acceptable risk. For more information on environmental decision making, the reader is referred to the author's companion book entitled *Global Environmental Policy*.[33]

Judging risk based on dramatic, catastrophic, and involuntary actions

Some factors tend to magnify apparent risk regardless of the outcome of a risk assessment. Involuntary hazards (the individual has no control over the event) are usually viewed as more serious than hazards faced by choice.[34] Thus, a comparison of the risks associated with mountain climbing or smoking, with the risk of a hazardous waste incinerator, may not be viewed as equivalent even if the incineration risk is much less. This is because the former are voluntary actions, under the control of the individual, whereas the latter risk is imposed by an outside party and not necessarily subject to the direct control of the individual.

Risks that are seen as potentially catastrophic, although unlikely, tend to be viewed as greater than risks from hazards that are more likely but would result in less serious or reversible outcomes. This is often true even though the overall risks (probability multiplied by the impact) of the two events are equivalent. It has also been observed that people pay more attention to new, dramatic, or unknown risks. This is also true of risks conveyed within the context of a personal story. As an example, the risk of a nuclear power plant may be viewed by the public as greater than the risk from a coal power plant, even though the risk of coal-fired plant emissions may be significantly more hazardous.

Risk also tends to be viewed as greater for hazards that impose a feeling of dread.[35] For instance, most people give proportionally more credence to a dramatic risk of dying from an airplane crash, than to the risk of dying from lung cancer due to smoking, even though the risk of the latter is much greater. Drama, symbolism, and identifiable victims, particularly children or celebrities, also make a risk more memorable. For example, a hazard to children is often judged significantly worse than a similar hazard to adults.[36]

Judgments of risk based on occupational, gender, and demographics differences

The population affected by a hazard is also important. Different groups often perceive risk differently. Technical specialists frequently view risks as less significant than do nontechnically trained persons.[37] For instance, the public often tends to be more concerned about very low levels of a toxic chemical than are toxicologists or physicists. In one study, nearly 100% of toxicologists, but only 70% of public respondents, agreed that a 1 in 10 million lifetime risk of cancer from exposure to a chemical is too small to worry about.[38]

The work environment is another factor in how people judge risk. Toxicologists who work for industry rate risks from chemicals significantly lower than toxicologists who work at universities.[39] For instance, managers who worked for British chemical companies had much lower perceptions of risk from chemical exposure than did members of the British Toxicological Society, who in turn judged risk lower than did members of the public.[40]

Men tend to rate risks lower than do women.[41] This difference in risk perception is not explained by differences in familiarity with scientific issues because female toxicologists have been shown to perceive chemical risks as higher than their male counterparts.[42] It is interesting that the gender difference in risk rating is seen only in white individuals. White men tend to rate risks less seriously than do black men, black women, and white women.[43]

The public also tends to be more concerned with risks involving many people at once as opposed to an individual.[44] Moreover, if an individual or organization imposing the risk is trusted by the community (e.g., established local company that hires a large part of the local workforce), the risk is often perceived to be less than if the risk is imposed by an unknown or outside party.

Judgments of risk based on social considerations

A risk viewed as unfairly distributed is often seen as larger than a risk that is fairly distributed.[45]

Finally, low-income and minority communities have become increasingly concerned about a disproportionate and unfair burden of environmental risk on their communities. Even a small risk may be viewed in the context of historical discrimination as contributing to an already unacceptable background of risk. The area of environmental justice (see author's companion text entitled *Global Environmental Policy*) is important because justice issues often involve hazards to minority and low-income communities.[46]

Risk communications

Risk is never a purely objective or scientific issue. Risk contains both scientific and social components that are subject to interpretation. As just witnessed, risk is routinely perceived differently, depending on characteristics of the hazard, the individual perceiving the risk, and its social context. Many scientists and technical experts who attempt to explain a risk fail to realize that the audience may perceive these same risks very differently. Failure to recognize this fact, and deal appropriately with such differences, can cause a significant risk communication failure.

Experience demonstrates that a successful risk communication strategy needs to address how the public perceives risk, how the media translate risk-based information, and how representatives and leaders can better relate risk information over a wide range of disciplines. In democratic societies, decision-making processes have increasingly involved the public as cooperative partners, often creating a perceived risk among technical specialists that the communication would be driven by non-experts.

This evolution has created a need for a systematic approach to risk communication in implementing public policy. Risk communication is an art requiring skill, knowledge, training, and adequate funding support. Effective risk communication involves both proper dissemination of the information, and communicating the complexities and uncertainties associated with the risk.

Framing

Framing is a fundamental problem with virtually every aspect of risk assessment.[47] The human mind can become overloaded, and people may compensate by taking "mental shortcuts." Among risk assessment analysts, this phenomenon may lead to discounting the risk of extreme events because the probability is considered too low to evaluate intuitively.

"Framing" refers to the observation that the context in which information is presented affects how risk communication is perceived. Studies indicate that a different framing of the same options can induce people to change their preferences among options.[48,49] This phenomenon is referred to as *preference reversal*. Consider the following example. A food distribution organization estimates that governments need to provide $500 million to avert a human disaster. Would a decision on whether to grant the

$500 million be different if the consequences of that decision were simply framed differently? Consider the following two statements:

- Spending $500 million will ensure that 75% of those affected by the famine will survive.
- Spending $500 million will still lead to the deaths of 25% of those affected by the famine.

Although these two statements are essentially equivalent, most people tend to choose the first one based on the way it has been framed.

One study of framing involved how people perceived lung cancer treatment. This study was based on the fact that medical statistics suggest that surgical treatment has a higher initial mortality rate, yet radiation has a higher 5-year mortality rate:

- Approximately 10% of surgery patients die during treatment, 32% will die 1 year following surgery, and 66 will die within 5 years.
- With respect to radiation treatment, 23% die within 1 year and 78 die within 5 years.

Yet, when people are given these mortality statistics, they tend to be evenly split between preferring surgery and radiation. However, when the same statistics are given in terms of overall life expectancies (6.1 years for surgery versus 4.7 years for radiation), there is an overwhelming preference for surgery.[50]

Another study found that when the issue of responsibility was removed from the way in which the question was framed, subjects were more likely to opt for vaccination versus assuming potential side effects of a vaccination; the factor of responsibility was removed by reframing the question: If you were the child, what decision would you like to see made?[51]

An unfortunate event that everyone agrees is inevitable can be ruled out of an analysis due to an unwillingness to admit that it is believed to be inevitable. Such human tendencies can affect even the most rigorous applications of the scientific method.

Finally, even the most rigorous risk assessments may be subject to peer pressure or *groupthink*—the acceptance of an obviously incorrect assessment simply because it is socially painful to disagree.

Anchoring and compression

Cognitive rules of thumb (heuristics) affect peoples' quantitative judgments of risk. Heuristics can affect estimates of risk in regular and predictable ways. As described below, use of these heuristics can result in biases in quantitative estimates of risk.

Risk estimates used for comparison, as well as the order in which they are presented, can affect how risks are perceived. "Anchoring" refers to the tendency to estimate the frequency of an event or risk on the basis of numbers presented for other events. Thus, if a person is told that 1,200 people a year die from electrocution and then is asked to estimate how many die from influenza, the response is likely to be lower than if the person is first told that over 40,000 people a year die in automobile accidents.

Events that draw media attention tend to be viewed as more likely. If, for instance, a particular risk has recently or often been reported in the popular press, people are more likely to overestimate its frequency or risk. Moreover, man-induced hazards also tend to be viewed as *riskier* than naturally occurring phenomena, even if the overall risk is equivalent.

Familiar hazards are generally seen as less risky compared with unfamiliar hazards.

"Compression" refers to the tendency to overestimate the frequency of risks that are uncommon and underestimate those that are common.

Reducing risk communication barriers

Communication barriers can frequently be traced to linguistic differences between scientists and laypersons. For example, scientists tend to approach the topic with the aim of educating people, explaining the scientific aspects of an issue, but not actively listening and responding to legitimate concerns voiced by the lay audience. In contrast, the general public often addresses risk communication in a legalistic or adversarial manner. Such differences in orientation can diminish the element of trust and credibility, leading to significantly increased controversy. Complicating this problem is evidence suggesting a growing distrust of experts—and of science in general.

Studies suggest that to be effective, risk communications must evoke a sense of personal relevance and trust in the recipient, and also demonstrate that the recipient can do something to reduce or control the risk. As outlined in Table 3.2, an effective risk communication program requires an understanding of a number of pertinent factors.

Addressing public concerns

Table 3.3 depicts the types of information that stakeholders concerned about health issues are likely to seek. A successful communication program depends on the ability to address these concerns.

Table 3.2 Factors Affecting Risk Communications

Nature of benefits:
 • Expected benefits associated with the risk
 • Who and how one benefits from the action
 • How the risk balances with the benefits
 • Magnitude of the benefits

Risk management issues:
 • Actions individuals may take to minimize their personal risk
 • Effectiveness of a specific option
 • Justification for selecting a specific course of action
 • Benefits of a specific option
 • Cost of managing the risk, and who pays for it
 • Residual risk remaining after risk mitigation measures are implemented
 • Action taken to control or manage the risk

Nature of risk:
 • Size and nature of the population at risk
 • Who is at the greatest risk
 • Urgency of the situation
 • Importance and characteristics of the hazard
 • Magnitude (severity) of the risk
 • Whether the risk is increasing or falling (trends)
 • Probability of exposure to the hazard
 • Distribution of exposure

Uncertainties in risk assessment:
 • Significance of uncertainties
 • Uncertainties in the data
 • Methods used to assess risk
 • Assumptions used
 • Sensitivity of the risk estimates to changes in assumptions

Table 3.3 Types of Information that Stakeholders Concerned about Health
Issues Are Likely to Seek

Mitigation measures
Nature of the risk
Severity of the risk
Degree of uncertainty
Who is potentially affected
Monitoring systems
Effective community communications
Demonstrated effort to reduce the risk and any uncertainties associated with it
Emergency response systems

Table 3.4 Seven Cardinal Rules of Risk Communication

1. Plan and evaluate the desired outcome of the communications effort (diverse projects, goals, audiences, and media all require different strategies).
2. Be honest, open, and sincere. Once lost, credibility is often difficult or impossible to re-establish.
3. Involve the public as a sincere partner. The ultimate goal should be to produce an informed public, not to defuse public concerns or replace actions.
4. Actively listen to the public's concerns. This will help establish trust, credibility, fairness, and empathy more than cold data and scientific statements.
5. Work with other parties. Conflicts among organizations can degrade communications with the public.
6. Work with the media to ensure accurate communications.
7. Speak clearly and with empathy, and recognize that some people will not be satisfied.

Developing a risk communications strategy

A risk communications strategy can be most effectively implemented through a two-stage approach:

Phase 1—Public Input: Stakeholders are queried to determine their risk concerns (i.e., how they perceive risk, and whom they trust).

Phase 2—Implementation: Risk communicators assess and determine a practical strategy, method, and communication channel (e.g., Internet, pamphlets, public forums or presentations, and newspaper articles) for implementing the communication campaign.

Covello and Allen have developed seven cardinal rules of risk communication, and they are noted in Table 3.4.[52]

Remember that the media are usually more interested in politics than in risk, controversy over the mundane, simplicity more than complexity, and in danger over safety. So actively engage the media to ensure accurate and fair reporting.

Communicating risk

People's perceptions of the magnitude of risk are influenced by factors other than numerical data. How a risk is explained and framed can have a significant effect on public opinion. Table 3.5 summarizes some general factors identified by Fischhoff et al. that influence the public's perception of risk.[53, 54] The field of *risk communication* (RC) is a multidisciplinary process of increasing importance in today's highly technological society. Public health officials use RC to give citizens necessary and appropriate

Table 3.5 Some Common Factors Influencing Risk Perception

- Risks that are under an individual's control are more acceptable than those that are not.
- Voluntary risks are more acceptable than those imposed upon an individual.
- Risks that have clear benefits are more acceptable than those with benefits that are less clear.
- Unfairly distributed risks are less acceptable than those deemed to be fairly distributed among a population.
- Familiar risks tend to be more acceptable than exotic ones.
- Risks that affect adults are viewed as more acceptable than those that affect children.
- Natural risks tend to be more acceptable than man-made ones.
- Risks generated by a trusted source tend to be more acceptable than those generated by an untrusted source.

information and to involve them in making decisions that affect them, such as where to build a toxic disposal facility.

Its applicability spans a spectrum of environmental policy and decision-making issues such as air pollution, hazardous waste sites, lead, pesticides, drinking water, and asbestos, drugs, and nanotechnology. Risk communication can also help promote changes in individual behavior such as in informing homeowners about the need to reduce carbon emissions.

Many risk issues that experts treat as scientific facts are, in truth, value judgments. They are what Fischoff has called "opinions of experts," as opposed to "expert opinions."[55] For its part, the public has become increasingly suspicious and distrustful of science. This does not imply that the public rejects good science, but rather that consumers are often very knowledgeable about the limits of science. They remember how often experts have been wrong in the past, how yesterday's orthodoxy is today's heresy, and how unsuspected risks have emerged from technologies whose proponents initially predicted nothing but great benefits. They also understand that scientists possess biases and judgments, just like everyone else; accordingly, they can be skeptical of arguments presented as good science that are, in fact, heavily laden in personal values.

Sandman has coined the term "outrage" to encompass many of the qualitative dimensions of risks.[56] As used by Sandman, a "hazard" is the quantitative, measurable aspect of a risk—how likely it is to kill you, while "outrage" encompasses all the attributes of a risk that determine how likely it is to worry or anger a person. Sandman points out that while the public may appear concerned with outrage at the expense of hazard, experts often ignore outrage, at their own peril. If the public feels its legitimate concerns are not being addressed, the outrage level will be greater than when the public feels listened to.

Perhaps the most important barrier to communication in food safety debates is mutual distrust between experts, and consumers. Distrust of the motives, attitudes, and beliefs of the other side makes it difficult to listen to, let alone accept and respond respectfully to, what the other side is trying to communicate.

Simplify language, not content

In communicating a complex issue such as environmental risk, it is easy to leave out information that seems to be overly technical. But risk communicators have shown that nearly any audience can understand most technical material as long as it is presented properly. For example, visuals and diagrams can be very useful in defining technical and medical or scientific jargon, including acronyms.

Dealing with uncertainty

When communicating risks, the results are not always definitive. Areas of uncertainty, such as how the data were gathered, analyzed, and how the results were interpreted should be described. This demonstrates that uncertainties are recognized, which helps to establish trust and credibility. The communicator should stress his or her expertise in the subject, which reinforces the image of leadership and the ability to handle the situation. It can also allay concerns and fears.

Recognize that safety is relative

Parties can communicate better if they remind each other that most technologies are neither safe nor unsafe, in absolute terms. Instead, the debate often wages not on whether something is safe, but whether it is safe enough. Such judgment requires either a balancing of risks versus benefits, or some comparison of the risk under consideration with other, similar risks that are judged either acceptable or unacceptable.

Defining the key safety question as "safe enough?" forces parties on both sides to discuss the value components of the decision, as well as what is known and not known on the scientific side. It is a major step toward more effective and clear communication.

Exercise caution when using risk comparisons

A risk can be put in perspective by comparing an unfamiliar risk to a familiar one. But use caution, as some types of comparisons can alienate the audience. Avoid comparing unrelated risks, such as smoking versus air contamination (the first is an individual's personal choice, the second is imposed). Moreover, there is a tendency among people to reject comparisons of unrelated risk.

Develop a key message

Key messages are of utmost importance that need to be communicated clearly, concisely, and to the point. Avoid deluging the audience—use no more than three messages at one time. Repeat key messages often to ensure they are not forgotten, misunderstood, or misinterpreted.

Accident analyses

An EIA analysis such as NEPA might need to address the effects of a potential accident so as to inform the decision maker and public about reasonably foreseeable adverse consequences associated with the proposal. The term "reasonably foreseeable" extends to events that might have catastrophic consequences, even if their probability of occurrence is low, provided that the analysis of the impacts is supported by credible scientific evidence, is not based on pure conjecture, and is within the rule of reason (40 CFR 1502.22).

With the exception of 40 CFR 1502.22, the US Council on Environmental Quality (CEQ) has not issued detailed guidance for addressing accident analyses in NEPA documents. This section provides guidance on performing an accident analysis that is generally applicable to the preparation of both environmental assessments (EAs) and environmental impact statements (EISs).[57]

For the purposes of NEPA, an accident can be viewed as an unplanned event or sequence of events that results in undesirable consequences. Accidents can be caused by equipment malfunction, human error, and natural phenomena.

Overview

An accident is an event or sequence of events that is not intended to happen, and indeed might not happen during the course of an operation. The probability that a given accident will occur within a given time frame, however, can be estimated. The probability of occurrence is expressed by a number between 0 (no chance of occurring) and 1 (certain to occur). Alternatively, instead of a probability of occurrence, one can specify the frequency of occurrence (e.g., once in 200 years, which also can be expressed as 0.005 times per year).

An accident scenario is the sequence of events, starting with an initiator, that triggers the accident. It is important to distinguish the probability (or frequency) of the accident initiator from that of the entire scenario; the latter quantity is of primary interest in NEPA accident analyses as it expresses the chance (or rate) that the environmental consequences will occur.

As used in this chapter, the environmental consequences of an accident are the effects on human health and the environment. In discussing an accident's effects on human health, it is both conventional and adequately informative to consider three categories of people: involved workers, noninvolved workers, and the general public. For each of these categories, effects should be evaluated for the maximally exposed individuals in these categories, and the collective harm to the entire population. This might involve identifying and quantifying, as appropriate, potential health effects (e.g., number of latent cancer fatalities).

In the context of analyzing accidents, the environment includes biota and environmental media, such as land and water, which can become contaminated as the result of an accident. The following guidance refers to effects on biota as ecological effects.

Consistent with the principle that impacts be discussed in proportion to their significance (§1502.2[b]), analysts should use a sliding-scale approach in determining whether an accident analysis is appropriate, as well as the degree of effort expended in performing such an analysis. Practitioners must apply professional judgment in determining the appropriate scope and analytical requirements. For example, practitioners need to determine the range and number of accident scenarios to consider, the level of analytical detail, and the degree of conservatism that should be applied. A sliding-scale approach is particularly useful in making these determinations (Table 3.6).

"Risk" can be used to express the general concept that an adverse effect could occur. As described earlier, in quantitative assessments, it is most commonly understood to refer to the numeric product of the probability and consequences.

Accident scenarios and probabilities

The following subsections provide guidance on addressing accident scenarios and probabilities.

Table 3.6 Factors to Consider in Applying a Sliding-Scale Approach to an Accident Analysis

- Severity of the potential accident impacts in terms of the estimated consequences
- Probability of occurrence and overall risk
- Context of the proposal (e.g., near a populated area versus a sparsely populated one)
- Degree of uncertainty regarding the analyses
- Level of technical controversy regarding potential impacts

Applying the rule of reason

One statutory basis under which US agencies may need to perform a risk assessment involves high-risk events. Based on case law, a NEPA analysis is governed by the rule of reason (see the author's companion book entitled *NEPA and Environmental Planning*).[58] The rule of reason is also interpreted to apply to the range of accident scenarios investigated in an EIS. Consistent with this rule, agencies are responsible for determining the appropriate range of alternatives to be evaluated. Under the rule of reason, an agency is not required to consider all possible accident scenarios or alternatives to a proposed action. Rather, the agency needs to consider "only a reasonable number of examples, covering the full spectrum of alternatives."

What constitutes a reasonable range of alternatives or accident scenarios depends on the nature of the proposal and the circumstances of each case. In general, the smaller the impact, the less extensive the investigation of alternatives is required. However, reviewing courts have generally insisted that an agency consider such alternatives as may partially or completely meet the proposal's underlying need. As a consequence, the scope of alternatives to be evaluated is a function of how narrowly or broadly the objective of its proposed action is viewed. For example, a major action involving a highly risky transportation of hazardous waste may require considering a full spectrum of alternatives (i.e., modes of transportation and alternative routes) that would adequately protect the human environment.

The rule of reason governs not only the range of alternatives to be considered, but also the extent to which they must be investigated. An agency's requisite consideration of alternatives should adequately articulate the reasons for the agency's choice and its rejection of alternatives. An agency is not required to select any particular alternative, and the examination of alternatives need not be exhaustive; however, it must be sufficient to demonstrate reasoned decision making. Therefore, an agency contemplating a major action such as transportation of hazardous waste would generally perform an appropriate risk assessment for each alternative (within the full spectrum of available and appropriate transportation mode alternatives) in the process of developing a well-reasoned decision.

Range of accident scenarios

Development of realistic accident scenarios that address a reasonable range of event probabilities and consequences is the key to an informative accident analysis. The set of accident scenarios considered should serve to inform the decision maker and the public of the overall accident risks associated with a proposal. As appropriate, accident scenarios should represent the range or spectrum of reasonably foreseeable accidents, which may include both low-probability/high-consequence accidents and high-er-probability/lower-consequence accidents.

Remember that the purpose of preparing an accident analysis in an EA is different from that in an EIS. In an EA, the purpose of an accident analysis is to determine whether a significant impact *could* result, requiring preparation of an EIS. Accordingly, the EA analysis normally focuses on the accident that could result in the maximum reasonable consequences.

If the accident analysis indicates that it is reasonable to conclude that a significant impact could result, an EIS must be prepared. Thus, where there is a potential for significant consequences, an EA normally focuses on the maximum reasonably foreseeable accident(s) that represent potential scenarios at the high-consequence end of the spectrum. In contrast, an EIS seeks to explore a range of different accidents and consequences that will assist the decision maker in discriminating and choosing among various alternatives, some or all of which involve significant impacts.

A maximum reasonably foreseeable accident is usually an accident with the most severe consequences that can reasonably be expected to occur for a given proposal. Such accidents tend to have lower probabilities of occurrence. Note that a maximum reasonably foreseeable accident is not the same as a worst-case accident, which almost always includes scenarios so remote or speculative that they are not reasonably foreseeable. Such events might be on the order of a comet colliding with a train transporting hazardous waste. Analysis of worst-case accidents is not required under NEPA.

An accident analysis does not necessarily end here. Accidents in the middle of the spectrum might also require evaluation, as they often contribute to, or even dominate, the overall risk assessment spectrum.

Equally important, a bounding approach that considers only the maximum reasonably foreseeable accident might not adequately represent the overall accident risks associated with the proposal. Further, an overly bounded analysis might prevent the decision maker from effectively, discriminating among alternatives and appropriate consideration of mitigation, because they can mask real differences among the alternatives.

Scenario probabilities

Accident scenarios can involve a series of events for which an initiating event is postulated. The initiating event would be followed by a sequence of other events or circumstances that result in adverse consequences. If these secondary events always occur when the initiator occurs (i.e., the secondary events have a probability of 1 given that the initiator occurs), then the probability (or frequency) of the entire accident scenario is that of the initiator. Otherwise, the scenario probability would be the product of the conditional probabilities of the individual events.

Risk

It is generally insufficient to simply present the reader with the risk of an accident (calculated by multiplying the probability of occurrence times the consequence). Presenting only the product of these two factors can mask their individual factors. Accordingly, risk should augment and not substitute for the presentations of both the probability of occurrence and the consequence of the accident.

Conservatism

Practitioners must exercise professional judgment in determining the appropriate degree of conservatism to apply. Preparers should consider the fundamental purposes of the analysis (e.g., purpose of an EA versus that of an EIS), the degree of uncertainty regarding the proposal and its potential impacts (see further discussion of uncertainty below), and the degree of technical controversy. In short, accident analyses should be realistic enough to be informative and technically defensible.

Scenarios based on pure conjecture should be avoided (40 CFR 1502.22). A method known as *compounded conservatism* involves bounding potential impacts, not by overestimating the effects or using error margins, but instead by using conservative assumptions; bounding is achieved by compounding these conservative assumptions.

Professional judgment must be exercised in compounding conservatism, as multiple conservative values can yield unrealistic results. For example, in air dispersion modeling, it is nearly always unrealistic to assume only extremely unfavorable meteorological conditions.

Thus, it is generally inappropriate to assume only the most severe conditions for an otherwise appropriate and credible accident scenario and then fail to analyze the scenario because, by using these conservative values, the overall probability is judged to be not reasonably foreseeable.

Accident consequences

Guidance for addressing the consequences of an accident is provided in the following subsections.

Uncertainty

A decision maker needs to understand the nature and extent of uncertainty in choosing among alternatives and considering potential mitigation measures. Where uncertainties preclude quantitative analysis, the unavailability of relevant information should be explicitly acknowledged. A NEPA document should describe the analysis used and the effect that the incomplete or unavailable information has on the ability to estimate the probabilities or consequences of reasonably foreseeable accidents (40 CFR 1502.22).

Based on the prevailing circumstances, practitioners can compensate for analytical uncertainty by using conservative or bounding approaches that tend to overestimate potential impacts. In other circumstances, such as where substantial uncertainty exists regarding the validity of estimates, a qualitative description may suffice. In all cases, however, the NEPA document should explain the nature and relevance of the uncertainty.

Regardless of whether a qualitative or quantitative analysis is performed, references supporting scenario probabilities, and other data and assumptions used in the accident analysis, should be provided.

Sabotage and terrorism

A NEPA document might need to address potential environmental impacts that could result from intentionally destructive acts (i.e., acts of sabotage or terrorism). While intentionally destructive acts are not accidents per se, they may still require consideration.

Analysis of such acts (fire, explosion, missile, or other impact force) poses a challenge because the potential number of scenarios is limitless and the likelihood of attack is unknowable. Nevertheless, the physical effects of such destructive acts are often similar to or bounded by the effects of other types of accidents. For this reason, where intentionally destructive acts are reasonably foreseeable, a qualitative or semi-quantitative discussion of the potential consequences may be sufficient and included as part of the accident analysis.

The following is an example of a qualitative discussion that might be appropriate for a hypothetical proposal involving a terrorist act against a truck transporting chlorine:

> Explosion of a bomb beneath the transportation truck or an attack by an armor-piercing weapon is possible. However, analysis shows that the consequences of such acts would be less than or equal to those associated with a maximum reasonably foreseeable transportation accident.

Noninvolved and involved workers

Impacts to involved workers should be evaluated as part of an accident analysis. Noninvolved workers would be within the vicinity of the proposed action, but not directly involved with action. Any potential impacts to noninvolved workers should generally be considered part of an accident analysis.

In some cases, a credible estimate of risk exposure to involved workers may involve more details about an accident than could reasonably be foreseen or meaningfully modeled. As a substitute, the effects might be described semi-quantitatively or qualitatively, based on the likely number of people who would be involved and the general character of the accident scenario. For example, a qualitative analysis might indicate "seven workers would normally be stationed in the room where the accident could occur. While a few such workers might escape the room in time to avoid being seriously harmed, several would likely die within hours from exposure to toxic substances, and the exposed survivors might have permanent debilitating injuries, such as persistent shortness of breath."

A more detailed, semi-quantitative or quantitative discussion might be necessary for analyzing proposals with substantially greater risks.

NEPA accident analysis and case law

Some key court rulings or pronouncements that can influence an assessment of accidents are summarized below.

Judicial review of scientific issues

In the United States, a court must generally defer to the expertise of a federal agency when assessing difficult issues involving a scientific or technical dispute, as long as the agency's determination is not *arbitrary and capricious*. When specialists express conflicting views, an agency has the discretion to rely on the reasonable opinions of its own qualified experts, even if a court might find contrary views more persuasive. Under this standard, an agency determination is merely required to have a rational basis (i.e., to be within a range of opinion generally accepted by the scientific community or justifiable in light of current scientific thought).

Risk assessment methodology   Use of an overall (probabilistic) risk assessment methodology, such as validated computer models, to calculate the risks of activities such as transportation of radioactive waste, will generally be sufficient to satisfy a court.

Cumulative risk   Risks from activities such as radioactive or chemical doses received from transporting waste may be immeasurably small. However, when people have been exposed repeatedly to this minimal dose e.g., from historic shipping campaigns, the cumulative dose should be included in the risk calculations, with an explanation regarding the inventory of chemical or radiation, the number of people potentially involved, and the projected health effects and risks.

Use of bounding values   Use of conservative estimates or bounding values for certain variables in risk assessment calculations (e.g., weather conditions, maximum dispersal of a toxic inventory, topography, and emergency response times) are generally sufficient to satisfy the requirements of NEPA. However, using bounding values tends to obscure, lessen, or smear differences among alternatives, making comparisons (alternatives) required by NEPA more difficult. Hence, their use should be limited to cases for which more accurate and detailed assessment is not practicable.

Low probability/high consequence accidents   Potential effects of low-probability accidents of high and beyond-design-basis consequences may need to be considered. In a *beyond-design-basis accident*, the accident event exceeds the design basis of the proposed facility or project. For example, the US Department of Energy (DOE) considers accidents with a probability of occurrence of 10^{-7} (1 in 10 million) or more per year as "maximum reasonably foreseeable accidents." Accidents having a smaller probability of occurrence rarely need to be considered. Such guidance may vary according to the federal agency, type of activity to be studied, type of hazard, severity of the hazard, and other factors.

Human error and sabotage   Some US courts have ruled that human error or sabotage should be considered in a risk assessment. Consider the chlorine transportation proposal. Information regarding the analysis of accidents as a result of human error (e.g., vehicle operation) can frequently be obtained from historic accident rates. To the extent that such information can be obtained (e.g., a probability of occurrence can be obtained from past historical accident data) and an accidental consequence (e.g., release of chlorine that would result from an accident) can be computed, such factors should be considered. Acts of sabotage or terrorism may also need to be considered in the risk assessment.

Endnotes

1. Goldstein, D., Demak, M., and Wartenberg, D., Risk to groundlings of death due to airplane accidents: a risk communication tool, *Risk Analysis*, 12, 339–341, 1992.
2. Alvarez, L. W., Alvarez, W., Asaro, F., and Michael, H. V., Extraterrestrial cause for the Cretaceous–Tertiary extinction, *Science*, 208, 1095–1108, 1980.
3. Chapman, C. R., and Momson, D., Impacts on the earth by asteroids and comets: assessing the hazard, *Nature*, 367, 33–39, 1994.
4. Morrison, D., and the Spaceguard Workshop, The Spaceguard Survey: Report of the NASA International Near-Earth-Object Detection Workshop (Jet Propulsion Laboratory, NASA Solar System Exploration Division Report), January 25, 1992.
5. Travis, C., et al., Cancer risk management: An overview of regulatory decision making. *Environmental Science and Technology*, 21, 415, 1987.

6. Covello, V. T., and Merkhofer, M. W., *Risk Assessment Methods*. New York: Plenum Press, 1994. 319 pp.
7. Groth, E., Communicating with consumers about food safety and risk issues. *Food Technology*, 45(5), 248–253, 1991.
8. Sandman, P. M., Risk communication: facing public outrage, *EPA Journal*, 13(9), 21–22, 1987; Slovic, P., Perception of risk, *Science*, 236, 280–285, 1987.
9. Hubbard, D., *The Failure of Risk Management: Why It's Broken and How to Fix It*, New York: John Wiley & Sons, 2009.
10. OHSAS 18001:2007
11. Landsburg, S., Is your life worth $10 million?, *Everyday Economics* (Slate), March 3, 2008. http://www.slate.com/id/2079475/ (retrieved March 17, 2008).
12. Montague, P., Making Good Decisions, http://www.gdrc.org/uem/eia/risk.html (accessed November 22, 2008).
13. National Research Council, *Risk Assessment in the Federal Government: Managing the Process*. Washington, DC: National Academy Press, 1983.
14. Fan, A., Howd, R., and Davis, B., Risk assessment of environmental chemicals, *Annual Reviews of Pharmacology and Toxicology*, 35, 341, 1995.
15. US Department of Energy. Vulnerability Assessment Methodology, Electric Power Infrastructure, 2002.
16. 40 CFR §1500.4(n) and §1506.4.
17. 40 CFR §1507.2.
18. US Office of Management and Budget and US Office of Science and Technology Policy, Memorandum on Updated Principles for Risk Analysis, September 19, 2007, www.whitehouse.gov/omb/memoranda.
19. Knight, F. H., *Risk, Uncertainty, and Profit*. Boston, MA: Houghton Mifflin Company, 1921.
20. Hubbard, D., *How to Measure Anything: Finding the Value of Intangibles in Business*, New York: John Wiley & Sons, 2007.
21. Covello, V. T., Risk communication: An emerging area of health communication research in Communication Yearbook 15 Ed. by Deetz, S., Newbury Park: Sage Publications, pp. 359–373, 1992; Covello, V. T., The perception of technological risks: A literature review. *Technological Forecasting and Social Change*, 23, 285–297, 1983.
22. Aronson, E., Wilson, T. D., and Akert, R. M., *Social Psychology, Media and Research Update, fourth edition* Upper Saddle River, NJ: Prentice Hall, p. 175, 2003.
23. Eccleston, C. H., *Global Environmental Policy: Principles, Concepts, and Practice*, Boca Raton, FL: CRC Press, 2010.
24. Festinger, L., *A Theory of Cognitive Dissonance*. Stanford, CA: Stanford University Press, 1957.
25. Wilde, G., (20001). *Target Risk 2: A New Psychology of Safety and Health*, Toronto: PDE Publications, 2001.
26. Thaler, R., Toward a positive theory of consumer choice, *Journal of Economic Behavior and Organization*, 1, 39–60, 1980.
27. Samuelson, W., and Zeckhauser, R., Status-quo bias in decision making, *Journal of Risk and Uncertainty*, 1, 1–59, 1988.
28. Fischhoff, B., Lichtenstein, S., Slovic, P., Derby, S. L., and Keeney, R. L., *Acceptable Risk*. New York: Cambridge University Press, 1981.

29. Meszaros, J. R., Asch, D. A., Baron, J., Hershey, J. C., Kunreuther, H., and Schwartz-Buzaglo, J., Cognitive processes and the decisions of some parents to forego vaccination for their children, *Journal of Clinical Epidemiology*, 49, 697–703, 1996.
30. Armor, D. A., and Taylor, S. E., When predictions fail: the dilemma of unrealistic optimism, in Gilovich, T., *Heuristics and Biases: The Psychology of Intuitive Judgment*. Cambridge, UK: Cambridge University Press, 2002.
31. A running debate between Kahneman, D., Lovallo, D., and Flyvbjerg, B., in the *Harvard Business Review* (2003).
32. A running debate between Kahneman D, Lovallo D., and Flyvbjerg B. in the *Harvard Business Review* (2003).
33. Eccleston, C. H., *Global Environmental Policy: Concepts, Principles, and Practice*, Boca Raton, FL: CRC Press, 2010.
34. Goldstein, B. D., and Gotsch, A. R., Risk communication. In Rosenstock, L., and Cullen, M. R., Eds., *Textbook of Clinical Occupational and Environmental Medicine*, Philadelphia, PA: WB Saunders Co., pp. 68–76, 1994.
35. Slovic, P., Fischoff, B., and Lichtenstein, S., Perceived risk: psychological factors and social implications, *Proceedings of the Royal. Society, London* A, 376: 17–34, 1981.
36. Hage, M. L., and Frazier, L. M., Reproductive risk communication: a clinical view. In Frazier, L. M., and Hage, M. L., Eds., *Reproductive Hazards of the Workplace*, New York: Van Nostrand Reinhold, pp. 71–86, 1998.
37. Slovic, P., Malmfors, T., Mertz, C. K., Neil, N., and Purchase, I. F., Evaluating chemical risks: Results of a survey of the British Toxicology Society, *Human & Experimental Toxicology*, 16, 289–304, 1997.
38. Neil, N., Malmfors, T., and Slovic, P., Intuitive toxicology: expert and lay judgments of chemical risks, *Toxicology and Pathology*, 22, 198–201, 1994.
39. Slovic, P., Malmfors, T., Mertz, C. K., Neil, N., and Purchase, I. F., Evaluating chemical risks: Results of a survey of the British Toxicology Society, *Human and Experimental Toxicology*, 16, 289–304, 1997.
40. Mertz, C. K., Slovic, P., and Purchase, I. F., Judgments of chemical risks: Comparisons among senior managers, toxicologists, and the public, *Risk Analysis*, 18, 391–404, 1998.
41. Barke, R. P., Jenkins-Smith, H., and Slovic, P., Risk perceptions of men and women scientists, *Social Science. Quarterly*, 78, 167–176, 1997.
42. Slovic, P., Malmfors, T., Mertz, C .K., Neil, N., and Purchase, I. F., Evaluating chemical risks: results of a survey of the British Toxicology Society, *Human and Experimental Toxicology*, 16, 289–304, 1997.
43. Flynn, J., Slovic, P., and Mertz, C. K., Gender, race, and perception of environmental health risks, *Risk Analysis*, 14, 1101–1108, 1994.
44. Wilson, R., Examples in risk-benefit analysis, *Chemtech*, 5, 604–607, 1975.
45. Alhakami, A. S., and Slovic, P. A., Psychological study of the inverse relationship between perceived risk and perceived benefit, *Risk Analysis*, 14, 1085–1096, 1994.
46. Eccleston, C. H., *NEPA and Environmental Planning: Tools, Techniques, and Approaches for Practitioners*, Boca Raton, FL: CRC Press, pp. 107–109, 2009.
47. Tversky, A., and Kahneman, D., The framing of decisions and the psychology of choice, *Science*, 211(4481), 453–458, 1981.
48. Tversky, A., and Kahneman, D., Availability: A heuristic for judging frequency and probability, *Cognitive Psychology*, 5, 207–232, 1973.

49. Lichtenstein, S., and Slovic, P., Reversals of preference between bids and choices in gambling decisions, *Journal of Experimental Psychology,* 89, 46–55, 1971.
50. McNeil, B. J., Pauker, S. J., Sox, H. C. Jr., et al., On the elicitation of preferences for alternative therapies, *New England Journal of Medicine,* 306, 1259–1262, 1982.
51. Baron, J., The effect of normative beliefs on anticipated emotions, *Journal of Personality and Social Psychology,* 63, 320–330, 1992.
52. Covello, V. T., and Allen, F., *Seven Cardinal Rules of Risk Communication,* Washington, DC: US Environmental Protection Agency, Office of Policy Analysis, 1988.
53. A Primer on Health Risk Communication Principles and Practices. Prepared by Lum, M. R., and Tinker, T. L., Washington, DC: US Department of Health and Human Services, Public Health Service, Agency for Toxic Substances and Disease Registry, 1994.
54. Slovic, P., Perception of risk, *Science,* 236, 280–285, 1987.
55. Fischoff, B., Risk: A Guide to Controversy. Appendix C, pp. 211–319, in NRC 1989, cited below.
56. Sandman, P., Risk communication: facing public outrage, *EPA Journal,* 13(9), 21–22, 1987.
57. Department of Energy, Analyzing Accidents under NEPA, draft guidance prepared by the Department of Energy's Office of NEPA Policy and Assistance, April 21, 2000.
58. Eccleston, C. H., *NEPA and Environmental Planning: Tools, Techniques, and Approaches for Practitioners,* Boca Raton, FL: CRC Press, pp. 271–273, 2009.

chapter four

Social impact assessment and environmental justice

> *When I heated my home with oil, I used an average of 800 gallons a year. I have found that I can keep comfortably warm for an entire winter with slightly over half that quantity of beer.*
>
> **—Dave Barry, post-petroleum guzzler**

By the 1960s it was becoming increasingly evident that altering the environment of the natural ecosystem can also affect society at large. Many early projects such as road construction had little or no social planning. Not surprisingly, many social disasters have occurred, resulting in profound social as well as ecological repercussions. For example, in 1973, following the decision to build the Alaskan pipeline from Prudhoe Bay to Valdez, local Inuit tribal leaders began to ask questions about how the pipeline would change their customs and culture.[1] For instance, how would the influx of construction workers affect the lifestyle of the local culture?

The modern concept of socioeconomic impact assessment grew out of the enactment of the US National Environmental Policy Act (NEPA) of 1969. This led to a new term—*social impact assessment* (SIA) or *socioeconomic impact assessment*. The fledging discipline of SIA grew out of a need to apply the fields of sociology and economics to predict the social effects of development projects subject to the NEPA.

This discipline grew throughout the 1970s. By the early 1980s, many US federal agencies had formalized environmental and social assessment procedures in agency regulations. In 1986, the World Bank included both environmental and social assessment procedures in its project planning procedures.

Defining SIA

"Social impacts" can be defined as

> *The consequences to human populations of any public or private actions that alter the ways in which people live, work, play, relate to one another, organize to meet*

> *their needs and generally cope as members of society.*
> *This term also includes cultural impacts involving*
> *changes to the norms, values, and beliefs that guide*
> *and rationalize their cognition of themselves and their*
> *society.*

In reality, the social aspect of the SIA has often been neglected, and the concept of an integrated socio-economic impact assessment has frequently been more of an economic assessment or one that emphasizes demographic changes. To ensure the consideration of a social component, "socioeconomic impact assessment" has begun to replace "socio-economic impact assessment." As used in this chapter, "SIA" will be understood to mean socioeconomic impact assessment, which emphasizes both the social and economic implications of a proposal.

The discipline of SIA has been broadly defined as the process of assessing or estimating the social and economic consequences that are likely to result from a proposed project or policy. As used in this chapter, the concept of social impacts includes both social and cultural consequences that result from either public or private actions that could affect the manner in which people live, work, and relate to one another. In contrast, cultural impacts refer to changes in the values, customs, or beliefs of individuals within a society.

As a process, SIA also has the potential to contribute greatly to non-environmental impact assessment planning processes. For example, in evaluating the introduction of a new health care system into native communities, New Zealand health professionals incorporated SIA into the process of evaluation alternatives for managing social change.

For practitioners implementing social policy decisions, SIA offers a formalized procedure for predicting the consequences, and managing potential social change. Consistent with the objectives of NEPA, the goal is to provide a process that allows decision and policymakers to make informed decisions between various alternatives. The general SIA process involves

1. Defining proposed projects or policies
2. Assessing potential impacts of alternatives, including the proposed action
3. Identifying and assessing alternatives and mitigation strategies to minimize potential socioeconomic impacts
4. Implementing mitigation measures
5. Developing monitoring programs to gauge the success of mitigation and identify unanticipated impacts

Benefits and considerations

SIAs are frequently viewed as expensive and unnecessary "fluff." While many countries have statutory requirements for performing SIAs, there is seldom a requirement for the SIA results to be seriously considered. Because SIAs are often not afforded serious consideration, measures for mitigating significant socioeconomic impacts are seldom implemented.

Simply experiencing a circumstance involving rapid change, even if the change is ultimately beneficial, is the cause of stress for many individuals. Frequently, the greatest social impact of a proposed policy or project is simply the stress resulting from uncertainty associated with living or working near a major development project and the uncertainty about the socioeconomic impacts that may result (crime, congestion, changes in the social complexion). Experience indicates that the simple act of engaging local community leaders and citizenry, and actively seeking their input in shaping the outcome of a proposal, can often alleviate significant fears by instilling a feeling of empowerment.

Not surprisingly, preparation of an SIA can often reduce opposition, and actually lower costs and expedite many projects over the long run. Experience has shown that local residents within potentially affected communities have frequently made substantial contributions to SIAs even when they did not have experience in related planning procedures. Communities should seriously consider preparation of an SIA because in many cases, the costs of rectifying social and environmental development impacts will eventually be borne by the public sector, not by the private entities that sponsored the proposal.

Moreover, the cost of hidden indirect and cumulative impacts can greatly exceed the cost of the more obvious direct impacts. Once a local culture is affected, it can be permanently and irreversibly affected. Consequently, it is important to assess, prevent, or mitigate impacts before they occur. For social impacts, particularly indirect effects, it can be difficult to prove damage that will satisfy a court. Moreover, many impacts cannot be mitigated or rectified after the fact, so legal compensation may not be a suitable option. Compensation from project proponents for damage or impact they may cause often only covers the direct but not the indirect impacts. This increases the need to assess both direct and indirect socioeconomic impacts before a final decision is made.

In general, the cost of preparing the SIA should be borne by the project proponent and not by the government or affected community.

SIA in environmental impact assessments

The intent of the NEPA is to promote a better understanding of project impacts that hopefully lead to wiser decisions through the implementation of alternatives and mitigation. Countless examples exist where an understanding of the environmental consequences have profoundly shaped the ultimate course of action, including sometimes selecting the no-action alternative.

The definition of the environment in an environmental impact assessment (EIA) process has generally been expanded with varying degrees of acceptance to include a socioeconomic component. However, the success with which an SIA has shaped project plans is another matter. There are surprisingly few documented cases where an SIA alone has directly and profoundly affected the ultimate decision-making process.

While the NEPA implementing regulations provide excellent direction for assessing environmental impacts, they provide virtually no guidance with respect to analyzing socioeconomic impacts. US courts have mandated that some degree of socio-economic assessment must be included in an environmental impact statement (EIS). Some federal agencies have also issued SIA direction in their guidelines and regulations. Since 1993, the US Council on Environmental Quality has explored ways to formally incorporate the SIA into its revised EIA regulations. The US Agency for International Development has also incorporated SIA-like procedures (e.g., social soundness analysis) into their project proposals. Some recent rulings by US courts have also upheld the need to incorporate aspects of SIA analyses into NEPA analyses.

While not benefiting from the same level of success as impact assessment of physical or natural resources in NEPA or the international EIA process, socioeconomic practitioners have reached a general consensus on how the SIA should be prepared. For example, the SIA discipline has forged the *Guidelines and Principles for Social Impact Assessment*.[2] This guidance is designed to complement the US NEPA and international EIA processes. It provides guidance for implementing public involvement, scoping, defining baseline conditions, evaluating alternatives, mitigation, and monitoring. Although the SIA process is frequently recognized, it is not always rigorously integrated into the NEPA and EIA processes.

Principles for socioeconomic impact assessment

Ten principles underlying the SIA process are depicted in Table 4.1.

Problems, authority, and conflicts in SIA

SIA seeks to identify how different sections of a community would be affected by a project and to investigate measures for reducing such impacts.

Table 4.1 Ten Principles Underlying the SIA Process

1. Public Involvement: Identify and involve all potentially affected stakeholders.
2. Data Collection: Develop a plan for gathering necessary baseline data and managing and effectively dealing with unknowns and gaps in the data.
3. Consider use of an adaptive management process (see companion text, *NEPA and Environmental Planning: Tools, Techniques, and Approaches for Practitioner*[17]).
4. Objective Analysis of Impacts: Clearly and objectively identify who will benefit or be negatively impacted.
5. Focusing on Significant Impacts: Focus on truly significant issues and public concerns. Emphasize impacts identified by potentially affected stakeholders.
6. Identifying Methods, Assumptions, and Defining Significance: Explain how the SIA was performed, the assumptions used, and how significance was evaluated.
7. Identifying Key Planning Issues: Identify key issues or problems that could be resolved with changes to the proposed action or alternatives.
8. Employing SIA Practitioners: Use professionals trained in relevant social and economic sciences and impact assessment methods.
9. Establishing a Monitoring Program: Manage uncertainty via a monitoring program.
10. Mitigation Program: Avoid or reduce adverse impacts by implementing cost-effective mitigation measures. Consider the use of an Adaptive Management Process (see the author's companion text, *NEPA and Environmental Planning*.[18]).

The SIA tends to focus on local concerns as opposed to issues of a broader context; but in so doing, SIA practitioners must also realize that broader concerns sometimes outweigh purely local concerns in decision making.

Rabel and Vanclay have introduced the following five problems that can profoundly affect the success of the SIA process.[3]

1. *Difficult questions often arise as to who holds the legitimate interests in the community?* Other questions involve how the affected community is defined and identified. Typically, project costs and benefits are not distributed equally across the community. One of the tasks of an SIA practitioner is to identify the stakeholders who stand to gain or to be adversely affected (winners and losers) by the proposal. The analysis frequently focuses on identifying and examining the social distribution of costs and benefits, usually in terms of social class and ethnic minority groups. The analysis also attempts to assess how the nature of the community could change.

At first, this task may appear straightforward but the concept of *community* is a fluid and often ambiguous concept. Consider an action requiring rezoning of a rural property. It may involve situations where newcomers (perhaps middle-aged professionals) have very different concerns from the established (e.g., predominantly farming) community. A second example involves local recreational areas, such as beaches, that are frequently subject to tourist development and inundation of new visitors.

As newcomers arrive, the original residents may be forced out of the community by rent or price increases. Local government boards may become dominated by socially and politically astute immigrants who establish new building codes and standards that may exclude some of the original residents. In recreational areas, for example, conflicts may erupt between the locals and other residents such as those who visit the area in the summer or winter periods to experience the natural surroundings. For example, if an ecologically significant forested area, zoned for logging, is rezoned as a protected area barring logging, the local community may experience social impacts in terms of lost jobs, long-term unemployment, or forced migration from their community to seek other work. Paradoxically, however, if logging continues unabated, the community could experience social impacts in the form of lost opportunities for future ecotourism and recreational industries.

Ironically, as each successive wave of newcomers arrives, they often either want no further development that would induce a next wave of newcomers; or alternatively, they may want to develop new income-producing activities that would profoundly affect the existing social and environmental setting.

In a stable community (where the rate of departing or incoming arrivals is low), it may be relatively easy to identify bona-fide community stakeholders. But if the rate of community growth is high with successive waves of immigration into the area, at any point in time, how are SIA practitioners expected to reach consensus on what the prevailing community views are? Whose views are entitled to be considered?

If a community is experiencing rapid growth, newcomers may bring very different values and attitudes. Concerns of the developers may also sharply conflict with those of the established community. This raises additional and perplexing questions. Where a community would experience profound growth patterns, should the newcomers be regarded as part of the community such that their concerns are included in any impact assessment? Should the concerns of various groups be given an equal or different weighting? Or are the newcomers part of the problem?

Other questions arise. NEPA expresses a goal of protecting the environment for future generations. Yet protection of future generations is an issue that is often not viewed as being pertinent to a standard SIA. Should future generations not yet born to a community be considered as part of the public? Whose interests should be considered paramount? Such questions tend to be emotional and politically charged.

2. *The role of community participation in the SIA process.* This question raises many issues about the right of local communities to determine their own destiny independent of nonlocal or outside influence. In an ideal world, community involvement should lead to increased appreciation of the impacts and provide measures for reducing them. Yet, this is often not the case. Suppose the public is opposed to a project, yet based on an objective analysis, the proposal is found to be beneficial. Or consider an opposing situation in which the public is in favor of the project, but an objective analysis indicates that the social and environmental impacts are likely to outweigh the benefits. How does one weigh the merits and deal with this situation?

Fears associated with controversial projects such as hazardous waste incineration frequently far exceed the actual risk. Such projects are particularly prone to distortion. Yet, the risks of more mundane activities involving everyday risks, such as those associated with the effects of smoking, sports, or road accidents, are often under-perceived. The public's perception of risk is often based on emotional responses, which does not necessarily correlate with the actual risk. Where the general public is opposed to a proposal, the perception of project risks may be greatly exaggerated as a result of fear, rumors, or deceptive advertisement.

Special interest groups often define problems and interpret SIA conclusions from their own point of view. Frequently, they attempt to use SIAs to their advantage, even if it means distorting the study's findings. Such individuals often reject the results of independent consultants whose findings contradict their particular agenda or biases.

The general community often does not know the truth about a proposed development project. Moreover, project proponents may manipulate or even distort public opinion though deceptive advertising, scare stories, or through unsupported promises of economic prosperity.

The nature of the public participation process can greatly affect the outcome and acceptance or rejection of the project. Public participation, regardless of how carefully undertaken, is not a substitute for a thorough SIA analysis. Unfortunately, public meetings frequently constitute neither effective participation nor representation. Because they often involve one-way information transferal, they may not be adequately participative. Nor are they necessarily representative because only certain groups may attend such meetings; the same groups that tend to benefit most from developments are frequently also the most likely to be represented.

3. *Ensuring an Unbiased Analysis.* Most SIAs are undertaken at the behest of a government body, community group, or project proponent. Unfortunately, each one of these entities tends to have a vested interest either for or against the project. When a consultant is engaged directly by developers, with no independent oversight, the consultants may provide an analysis favoring the view of those who commissioned the study (often a pro-development line). Contrary or critical issues are often afforded insufficient attention. Unless an SIA is undertaken by independent professionals, the conclusions may be questionable.

4. *Gauging the impacts.* It is not uncommon to find that certain impacts may be perceived as negative by some members of the community, and as positive by others. This is because the impacts are subject to the value judgments of individuals. Consider the impacts of siting a new railroad maintenance yard in a rural community that involves relocation of a transient workforce and their families to that community. Some of the community may view this as a negative experience and may be concerned about the loss of community character and their personal safety. Other members may believe that the community was too narrow minded to begin with, and thus the infusion of a largely different type of people might be good because it will broaden mental attitudes. Moreover, some individuals may consider an impact as a mild inconvenience, while others may believe it will greatly change the local character of the community. The SIA cannot judge the impact; it merely reports how different segments of a community are likely to react and view the consequences. As is true for assessing impacts to natural and physical resources, impacts to socioeconomic resources are not always measurable using one or more readily quantifiable metrics. All impact assessment in the context of NEPA includes qualitative and quantifiable elements and requires sound professional judgment by the assessor.

5. *The decision-making process.* An SIA alone, as is true for EIA processes such as NEPA EISs or environmental assessments (EAs) in general, cannot make definitive decisions about whether a project should proceed. Decisions about whether a project should proceed may require consideration of both environmental and non-environmental factors. Properly orchestrated, an SIA, like other elements of an EIA can provide valuable information for reaching an informed decision and facilitating public discussion that may influence the ultimate decision.

Preparing the SIA

This section describes the general process followed in preparing an SIA. Emphasis is placed on integrating the SIA process with the NEPA planning process. The SIA is most commonly included as the socioeconomic sections of the EIA document, such as an EIS or EA. For some especially complex or controversial projects, the proponent agency may prepare a stand-alone SIA that is referenced in the EIA. However, the principles outlined in this section can be adopted with little modification to virtually any international EIA process similar to NEPA.

Socioeconomic impact assessment and NEPA

NEPA requires that federal agencies first prepare an EIS to investigate the impacts of reasonable alternatives and mitigation measures before they can take actions significantly affecting the quality of the *human environment*. Preparing an EIS requires the integrated use of the social sciences. The social science element of an EIS is referred to as the social or socioeconomic impact assessments, or simply SIAs. EISs are intended to provide decision makers with a full disclosure of the positive and negative effects of a proposal.

The US NEPA regulations found in 40 Code of Federal Regulations (CFR) Parts 1500–1508, state that the "human environment" is to be "interpreted comprehensively" to include "the natural and physical environment and the relationship of people with that environment" (40 CFR 1508.14). In addition to environmental impacts, an EIS must also investigate "aesthetic, historic, cultural, economic, social, or health" effects, "whether direct, indirect, or cumulative" (40 CFR 1508.8).

The regulations contain another key provision: "... economic or social effects are not intended by themselves to require preparation of an environmental impact statement" (40 CFR 8.14). However, where an EIS is prepared and economic or social and natural or physical environmental effects are interrelated, then the environmental impact statement will discuss all of these effects on the human environment (40 CFR 1508.14). Thus, an EIS is not required for a proposal that has a significant economic impact, but whose environmental impacts are nonsignificant.

The methodologies for assessing social impacts are numerous. Analysis of impacts on communities should be based on conceptual relationships developed from theory and peer-reviewed research, supported by data collected using valid methods, and subject to empirical verification.

Interrelation of SIA and impact assessment
for other environmental resources

SIA should not be viewed as some completely separated element of an EIA process such as NEPA, to be completed in a vacuum as some appendage to

Table 4.2 Interrelation of SIA and Topics Traditionally Associated with NEPA

Resource	Relation to SIA
Air quality	The quality of life for area residents is affected by air quality. Reduced air quality may discourage people from moving to an area and therefore discourage investment by businesses in that area.
Cultural (archaeological and historic) resources	Tourism opportunities, and the economic development they engender, can be limited by damage to culturally significant sites and properties on the National Register of Historic Places; the historic character of an area can be an important factor in its social identify and cohesion.
Ecology	Tourism and recreational opportunities, and the economic development they engender, can be limited by losses of natural habitats and wildlife.
Land use	The availability of land can affect development and investment opportunities in an area. Actions inconsistent with zoning laws or comprehensive planning objectives can reduce the attractiveness of an area to new residents and businesses.
Water quality	The quality of life for area residents is affected by the availability of clean surface and groundwater. Reduced water quality may discourage people from moving to an area and therefore discourage investment by businesses in that area. Tourism opportunities, and the economic development they engender, can be limited by degradation of rivers, streams, lakes, and other waters used for recreation.

the environmental impact assessment process. Socioeconomic issues should be a part and parcel of the human environment, closely interrelated with the physical and natural resources traditionally associated with the NEPA process. Table 4.2 illustrates the interrelated character of SIA and some of the more commonly associated elements of NEPA analysis.

Native Americans

Most US federal agencies have established government-to-government relationships with Native American tribes. Whenever a NEPA proposal has the potential to impact people living on a reservation, the regulations grant special status to tribes, particularly with respect to consultations, participation in the formulation of issues, and submitting comments on draft EISs.

In the SIA process, Native American rights have been expanded under the American Indian Religious Freedom Act (PL 95-341) and the

Native American Graves Protection and Repatriation Act of 1990. These groups are to be consulted whenever a federal action affects any of their culture's resources on or off current reservation lands.

Just as the biological sections of EISs devote particular attention to threatened or endangered species, the socioeconomic sections of EISs need to pay particular attention to the impacts on vulnerable segments of the human population: the poor, elderly, adolescents, unemployed, and minority groups.

In some agencies, public involvement tends to be synonymous with the SIA process. This results in a misleading belief that because they have achieved a public involvement they have satisfied the SIA function. The SIA addresses a particular category of resources, not a particular procedural element of the EIA process. Public involvement is simply a component of the SIA process, as it is for the assessment of impacts to natural or physical resources.

Evaluating the three stages of a project

An EIA process such as a NEPA analysis needs to evaluate probable undesirable social effects of a development action before they occur in order to support choices among alternatives and make recommendations for mitigation. As socioeconomic impacts are identified and evaluated, recommendations for mitigating actions can be made. For instance, a freeway extension facilitates residential growth, which leads to increased traffic and air pollution, creation of new schools, retail centers, and other services, as well as a decline of the downtown neighborhood. The analysis of alternatives and mitigation measures offers decision makers an opportunity to help rectify such adverse impacts.

Forecasted socioeconomic impacts represent the difference between the future with the project and a future without the project on humans and their communities. Because the future cannot be seen, analysts often look at communities that have been affected by similar projects in the past.

Information about the community or geographic area of study is frequently available both before and after the event to help in forecasting potential changes. One method of describing socioeconomic changes is to describe one or more (perhaps a series of) "snapshots" over time as the development unfolds. Socioeconomic impacts represent the changes taking place between these snapshots.

Most proposals go through a series of stages, starting with implementation and construction, and carrying through to operation and maintenance. Socioeconomic impacts will be different for each stage. A three-stage process is described below.[4] To the extent practical (particularly with respect to planning and evaluating Stage 3), the SIA should evaluate each stage of a typical proposal's life cycle.

Stage 1: Construction   The first stage—construction or implementation—begins when a decision is made to proceed. For typical construction projects, this involves clearing land, building access roads, developing utilities, etc. An influx of construction workers occurs, requiring lodging, transportation, and public services. Local population levels can increase rapidly, overwhelming support services. Equally disturbing can be the sudden loss of jobs and out-migration as construction activities reach completion. Displacement and relocation of people, if necessary, occurs during this phase.

Locally affected communities, particularly smaller ones, may have difficulties responding to increased demands on housing, school, health facilities, and other services. Additional stress may be created by sudden increases in the prices of housing and local services, or resentment between newcomers and long-time residents. Construction activities may change the traditional community fabric, replacing accepted behaviors.

Stage 2: Operation/maintenance   The second stage—operation/maintenance—generally occurs as the construction or implementation phase is nearing completion and continues through the operating stage of the project. This phase frequently requires fewer workers than the construction or implementation stage. If operations continue at a relatively stable level, the effects are often viewed as beneficial. Areas seeking development often focus on this stage because of the long-term economic benefits. Over the course of this stage, communities often adapt to the new socioeconomic conditions, and the positive expectations of a stable population, a quality infrastructure, and employment opportunities are realized.

Stage 3: Abandonment/decommissioning   The third and final stage—abandonment/decommissioning—typically begins as the project nears completion. The social impacts of decommissioning may begin when an announcement is made to cease or shut down the project. This may lead to loss of the economic base as supporting businesses close their doors. However, sometimes an influx of workers, similar to that experienced during the construction stage, occurs if a significant environmental clean-up is required. In such cases, the clean-up becomes its own separate project, with its own three-phased project trajectory.

Identification of socioeconomic assessment variables

Socioeconomic assessment variables are attributes that cause changes in the characteristics of human population, communities, and social relationships. A list of social variables that may need to be assessed include

- Cultural norms and values
- Political and social resources

- Community and institutional structures
- Individual and family changes
- Community resources
- Population characteristics

Sources of standard socioeconomic data   Socioeconomic data can be obtained from standard sources such as the US Census, state economic development agencies, local government agencies or Chamber of Commerce records, and private organizations that operate as data brokers. Some private institutions post data on their Websites. In some cases, this data is provided free of charge, often on Websites; some is provided for a nominal fee; and in still other cases, data can be purchased for a substantial fee. In the latter situation, the data have usually been processed in some way by a third party.

Important sources of socioeconomic data such as US Census data are typically summarized at different geographic levels: national, state, county, census tracts, and block group and block. Block level represents aggregate data for entire city blocks or an area delineated by the Bureau of the Census. Blocks are further aggregated into block groups, which comprise census tracts, perhaps the most familiar summary level for census data. Data are also generally summarized for selected political subdivisions and places.

Generalized socioeconomic impact assessment process

In general, a socioeconomic assessment process should contain the seven steps outlined below. This sequence is patterned after the EIA process outlined in the NEPA regulations, applicable to any resource addressed in the context of NEPA. The SIA practitioner should

- Focus on significant impacts
- Provide quantification where feasible and appropriate
- Present the social impacts in a manner that can be clearly understood by the decision maker and stakeholders

Step 1: Public involvement
Identify all potentially affected groups, at the early stage in the planning process. Potential groups include those who live nearby and those who will see, hear, smell, or in other ways experience effects. Other parties may include those affected by the influx of seasonal residents who may have to pay higher prices for food or rent, or pay higher taxes to cover the cost of expanded community services. Once identified, representatives from these groups may be interviewed to determine potential concern.

*Step 2: Establishing the baseline of human
environment and conditions*

Baseline conditions are the existing and past trends associated with the potentially affected human environment. The level of effort devoted to the description of the human environment should be commensurate with the size, cost, and degree of expected impacts of the proposed action. A geographical area is identified along with the distribution of special populations at risk. Baseline conditions might include

- Relationships with the physical environment
- Historical background (including initial settlement and subsequent population shifts)
- Political and social resources (including the distribution of authority)
- Cultural, social, and psychological conditions and attitudes
- Population characteristics
- Demographics of relevant groups (including stakeholders, and sensitive groups)

Step 3: Scoping

The scoping process is used to sharply focus the analysis on key impacts, actions, alternatives, and mitigation that will be evaluated in the SIA statement. Following initial scoping, relevant criteria for assessing significant impacts comparable to those spelled out in the regulations (§1508.27) should be established. Potential factors include

- Degree of social disruption or relocation
- Degree of economic impact
- Number of people, including indigenous populations that will be affected
- Duration of impacts (long-term versus short-term)
- Probability of the event occurring
- Uncertainty over possible effects
- Presence or absence of controversy over the issue
- Value of benefits and costs to impacted groups (intensity of impacts)
- Extent that the impact is reversible or can be mitigated
- Relevance to present and future policy decisions

Alternatives identification. The next step involves describing the proposal in sufficient detail to identify data requirements for performing the SIA:

- Institutional resources
- Incomes
- Facility description

- Land requirements
- Locations
- Ancillary facilities (roads, sewer, and water lines)
- Needs of workforce
- Construction schedule
- Workforce size (construction and operation)

Step 4: Impact investigation

In general, there is consensus on the types of impacts that need to be considered (social, cultural, demographic, economic, and possibly psychological and political impacts). Probable social impacts are formulated in terms of predicted conditions:

- Without the actions
- Predicted conditions with the actions
- Predicted impacts that can be interpreted as the differences between the future with and without the proposed action

Investigation of the probable impacts involves six major sources of information:

- Interviews
- Literature
- Data from project proponents
- Records of previous experience with similar actions
- Census and vital statistics
- Field research

Methods of projecting future changes lie at the heart of the SIA process. Some of the methods available include:

- *Linear trend modeling:* Taking an existing trend and simply projecting the same range of change into the future. This method can be useful in defining anticipated impacts under the no-action alternative.
- *Expert testimony:* Experts can be asked to present scenarios and assess their implications.
- *Scenario evaluation:* Evaluating postulated cases or scenarios.
- *Population multiplier methods:* An increase in the population implies a designated change (multiples) of other variable such as jobs or housing units.
- *Numerical computer modeling:* Numerical computer models are used to simulate or forecast socioeconomic changes, particularly in terms of demographics and economics.

As noted above, many techniques are employed in forecasting socio-economic impacts (trend analyses, econometric studies, and informal surveys). In using trend analyses or economic studies, the time span of the study must be long enough so that inferences can be made about the magnitude and direction of the coefficients of the variables. For example, while population statistics are widely considered in transportation impact studies, relating the indirect population changes to a specific project might be difficult.

Expert knowledge enlarges the knowledge base; it is useful in judging how the study case is likely to deviate from the typical patterns.

The opinions of individuals and groups toward the proposed change should also be assessed. Surveys provide an important means of verifying the analysis.

Step 5: Forecasting impacts

Indirect impacts occur either later than the direct impact or farther away. Cumulative impacts result from the incremental impacts of an action "added" to other past, present, and reasonably foreseeable future actions regardless of which agency or person undertakes them (see 40 CFR 1508.7). While they are more difficult to estimate than direct impacts, indirect and cumulative impacts must be addressed in NEPA analyses.

The analyst first determines how the proposal will affect the site and surrounding community. Once the direct impacts have been estimated, the assessors determine how the affected people will respond in terms of actions and attitude. Attitudes before project implementation are often a predictor of their attitudes afterward, although some studies show that fears are often exaggerated; experience also suggests that expected (frequently promised) benefits often fail to meet expectations.

How actions affect groups can often be estimated using comparable cases and through interviews with affected people. The ability to demonstrate to potentially affected people that significant impacts are being addressed in the analysis can be critical to a successful process.

Step 6: Assessing alternatives and mitigation

As necessary, alternatives may be reshaped, new alternatives are developed, and in particular, mitigation measures are developed to address potentially significant socioeconomic impacts. The number of iterations needed to complete this step will depend on time, funding, and the magnitude of the project or policy changes. Expert judgment and scenarios are helpful in developing alternations.

Mitigation   Under NEPA, mitigation includes avoiding the impact by not taking or modifying an action; minimizing, rectifying, or reducing the impacts through the design or operation of the project or policy; or compensating for the impact by providing substitute facilities, resources, or opportunities (see 40 CFR 1508.20).

Ideally, mitigation measures are built into the selected alternative, but it is appropriate to identify mitigation measures even if they are not immediately adopted or if they would be the responsibility of another person or government unit.

Two principal types of mitigation—avoidance and impact minimization—can apply to the project itself or to the community or the impacted region as a whole. For instance, a project may be revised to avoid or minimize adverse social impacts (e.g., extend the construction period to minimize immigration), or the community may take steps to attenuate, if not avoid, such effects. The first step in evaluating potential mitigation is to determine whether the proponent could modify the project to avoid the adverse effects. For example, a road that displaces a farm might be rerouted.

One of the benefits of assessing irresolvable social impacts is that a method for compensating individuals and the community for unavoidable impacts may be identified. Such compensation may spell the difference between project success and failure.

Step 7: Monitoring

Monitoring is an important aspect that should be seriously considered for all proposals. As appropriate, a monitoring program should be developed, capable of identifying deviations from the projected impacts and any important unanticipated impacts. A monitoring plan should track the project development and compare real impacts with projected ones.

Monitoring programs are particularly valuable for proposals that lack detailed information or involve a degree of variability or uncertainty. Where monitoring procedures cannot be adequately implemented, any mitigation agreements should acknowledge the uncertainty faced in implementing the decision.

Commonly encountered problems

There is a lack of recognition of the complexity of socioeconomic problems, issues, and acceptance by local communities. The physical sciences generally tend to present well-defined problems for which tangible solutions can be developed. This success leads to the belief that social issues are similarly well defined, which leads to an expectation that SIA statements will deliver clear statements of social impacts and successful mitigation programs. In reality, SIAs are often simply appended to an EIA statement. Little attempt is made to integrate and interpret the findings collectively or comprehensively. Such analyses are often reductionist and lack a holistic understanding of a proposal and its potential impacts.

Inexperienced consultants

Some officials naively assume that development is always good and that there are no social (and sometimes no environmental) consequences; there may be little recognition of the need for professional expertise to assess social impacts. Because little credence is given to socioeconomic implications, it may be assumed that anyone can determine such consequences. There is no registration of suitably qualified SIA practitioners, and some unscrupulous consultants may claim to possess such expertise when they do not. It is not uncommon to find that an SIA has been developed by an unqualified consultant who is not trained in SIA and the underlying economic and social sciences.

Scope, scale, and thresholds

SIAs need to address the scope and scale of the project. The social effects of some developments can be extremely dispersed from the original development site. The issue of the extent to which the analysis should be applied to larger geographical regions, as opposed to the immediate project vicinity, needs to be defined. But, as the spatial or temporal (time frame) scale expands, it becomes increasingly more difficult to assess the socioeconomic significance.

Most communities have a basic resilience that can accommodate a certain degree of growth or change. Impacts become increasingly apparent as the number or extent of changes exceeds a certain threshold. Prudent planning is necessary to ensure that such thresholds are not exceeded.

Mitigation

An SIA is all too often viewed as a discrete statement of impacts, not as a process that seeks to identify mitigation, or as a process that fosters good decisions. Consequently, approval might be denied to projects that would otherwise be acceptable provided certain mitigation measures were instituted. Conversely, projects are sometimes approved, with compensation paid, despite the fact that the compensation might not have been necessary—if only appropriate planning and mitigation had been used to avoid the impact in the first place.

The SIA makes it greatest contribution through its ability to identify and assess mitigation. Many possible impacts can be easily avoided by simple and cost-effective mitigation strategies that can turn development projects with negative social impacts into projects with positive impacts, at least for much of a community. Such an approach is recognized in certain industries, particularly the mining industry, where through the use of an SIA and community development consultants, practical social strategies and social design concepts have had a profound influence on community well-being.

Not only should mitigation measures be assessed, but appropriate political procedures may also need to be established to determine who is responsible for mitigation and monitoring.

Spirituality

One aspect of culture is spirituality, which is an integral aspect in many indigenous cultures. Many anthropologists argue that spirituality applies to all cultures, even if it is manifested in different forms that people may not recognize as spiritual. For many groups, particularly indigenous peoples, specific religious and spiritual issues should be considered. Yet, the reference to and the use of the "spiritual" in assessing human interaction with ecosystems is a surprising revelation for many; it often runs counter to traditional Western thought. Concepts such as attachment to the land or identification with a sacred location are difficult to quantify and easily discredited; yet such issues may be among the most important factors in determining project success and probable acceptance by local populations. Success may hinge on providing appropriate attention and cultural sensitivity to this issue.

Environmental justice

Environmental justice (EJ) has become a topic of special interest in recent years. Accordingly, this section provides practitioners with guidance on incorporating EJ considerations into both the preparation of environmental assessments (EAs) and EISs. This direction is based principally on a presidential executive order, and guidance developed by the President's Council of Environmental Quality, the US Department of Energy, and the US Environmental Protection Agency (EPA).[5]

The EPA defines EJ as[6]

> The *fair treatment* and meaningful involvement of all people regardless of race, color, national origin, or income with respect to the development, implementation, and enforcement of environmental laws, regulations, and policies. Fair treatment means that no group of people, including racial, ethnic, or socioeconomic group, should bear a *disproportionate share of the negative environmental consequences* resulting from industrial, municipal, and commercial operations or the execution of federal, state, local, and tribal programs and policies (emphasis added).

Implemented without prudence, EJ considerations can easily get out of hand, significantly adding to what may already be a costly and resource-intensive environmental process. Emphasis is therefore placed on providing the reader with a practical and balanced approach for addressing EJ.

As is recommended by the US Council on Environmental Quality (CEQ) for various environmental considerations under NEPA, a sliding-scale approach should be applied in determining the level of effort most appropriate for addressing EJ; that is to say, tasks such as identifying populations, assessing impacts, and enhancing participation should be performed commensurate with the potential for sustaining dispropor-tionately significant impacts

Under a sliding-scale approach, the depth and attention given to the analysis varies with the potential significance of impacts on minor-ity or low-income populations. The analysis may be either qualitative or quantitative.

Background

EJ began surfacing around 1994. Some of the principal direction and guid-ance documents that have since been issued are outlined below.

Executive Order 12898

In February 1994, President Clinton signed Executive Order 12898, *Federal Actions to Address Environmental Justice in Minority Populations and Low-Income Populations.*[7] It requires that each federal agency

> *...make achieving environmental justice part of its mis-sion by identifying and addressing, as appropriate, disproportionately high and adverse human health or environmental effects of its programs, policies, and activi-ties on minority and low-income populations.*

A presidential memorandum accompanying this executive order directs federal agencies to:

> *...analyze the environmental effects ... of federal actions, including effects on minority communities and low-in-come communities, when such analysis is required by the National Environmental Policy Act.*

Federal agencies are also instructed to:

> *...provide opportunities for community input in the NEPA process, including identifying potential effects*

> *and mitigation measures in consultation with affected*
> *communities and improving the accessibility of meetings,*
> *crucial documents, and notices.*

Council on Environmental Quality guidance

In 1997, the US Council on Environmental Quality (CEQ) issued *Environmental Justice Guidance under the National Environmental Policy Act*. This document provides direction on incorporating EJ into the NEPA process.[8] It states that the presidential executive order signed in 1994 (E.O. 12898) does not change prevailing legal thresholds and statutory interpretations under NEPA and existing case law. However, it emphasizes that agency consideration of impacts on minority or low-income populations can identify disproportionately high and adverse impacts that are significant and that might otherwise be overlooked.

The document goes on to point out that EJ issues encompass a broad range of impacts covered by NEPA, including impacts on the natural or physical environment and related social, cultural, and economic impacts. This guidance also acknowledges that EJ issues can arise at any step in the NEPA process, and agencies should consider these issues at every step of the process, as appropriate.

This guidance states that environmental impacts to minority and low-income populations do not have a different threshold for significance from other impacts, but specific consideration of impacts on minority and low-income populations can identify "disproportionately high and adverse human health or environmental effects that are significant and that otherwise would be overlooked."

Environmental Protection Agency guidance

In 1998, the EPA issued *Guidance for Incorporating Environmental Justice Concerns in EPA's NEPA Compliance Analyses*.[9] This guidance applies to NEPA reviews conducted by the EPA.

Soon thereafter, the EPA issued *EPA Guidance for Consideration of Environmental Justice in Clean Air Act Section 309 Reviews*.[10] It applies to EPA reviews (under Section 309 of the Clean Air Act) of EISs prepared by other federal agencies.

Analyzing environmental justice impacts

Agencies are instructed to evaluate proposals for their potential to produce disproportionately high and adverse human health or environmental effects.

Practitioners should note cultural differences among stakeholders regarding what constitutes an impact or the severity of an impact. For example, a Native American tribe might regard an act that provides the general public with access to a particular mountain as a desecration of its sacred site. Agency officials should also recognize that risk perceptions can vary widely, and commenters might disagree with the agency's underlying assumptions concerning risk factors.

Factors used in determining a disproportionately high and adverse impact

The CEQ document entitled *Environmental Justice Guidance under the National Environmental Policy Act* presents factors to consider when judging the importance of *disproportionately high and adverse* health and environmental impacts on minority and low-income populations (see Table 4.3).

Table 4.3 Factors Useful in Judging High and Adverse Impacts

Disproportionately high and adverse human health effects:
Agencies should consider the following three factors, to the extent practicable:
- Whether the health effects, which might be measured in risks and rates, are significant (as employed by NEPA), or above generally accepted norms. Adverse health effects may include bodily impairment, infirmity, illness, or death.
- Whether the risk or rate of hazard exposure by a minority population, low-income population, or tribe to an environmental hazard is significant (with respect to NEPA) and appreciably exceeds or is likely to appreciably exceed the risk or rate to the general population or other appropriate comparison group.
- Whether health effects occur in a minority population, low-income population, or tribe affected by cumulative or multiple adverse exposures to environmental hazards.

Disproportionately high and adverse environmental effects:
Agencies should consider the following three factors, to the extent practicable:
- Whether there is or will be an impact on the natural or physical environment that significantly (with respect to NEPA) and adversely affects a minority population, low-income population, or Indian tribe. Such effects may include ecological, cultural, human health, economic, or social impacts on minority communities, low-income communities, or tribes when those impacts are interrelated to impacts on the natural or physical environment.
- Whether environmental effects are significant (with respect to NEPA) and are or may be having an adverse impact on minority populations, low-income populations, or tribes that appreciably exceeds or is likely to appreciably exceed that on the general population or other appropriate comparison group.
- Whether the environmental effects occur or would occur in a minority population, low-income population, or tribe affected by cumulative or multiple adverse exposures from environmental hazards.

It is important to note that with respect to NEPA, *economic or social effects are not alone considered significant* (i.e., requiring preparation of an EIS). However, when an environmental impact statement is prepared, and economic or social, and natural or physical environmental effects are interrelated, then the EIS is to discuss all these effects on the human environment (§1508.14).

Evaluating high and adverse impacts   On completing the analysis of impacts on the general population, analysts should determine, consistent with CEQ guidance, whether any impacts on a minority or low-income population has the potential to be disproportionately high and adverse.[11] A two-step approach is warranted. Specifically, practitioners must judge

1. Whether the impacts on a minority or low-income population would be high and adverse (e.g., significant) within the meaning of NEPA (i.e., high and adverse)
2. Whether any of these high and adverse impacts would disproportionately affect a minority or low-income population relative to the general population

Impacts to a minority or low-income population considered to have a potential to be disproportionately high and adverse are analyzed. As with the analysis of impacts to the general population, a sliding-scale approach should be utilized in determining the level of review necessary to make judgments regarding the significance of impacts on minority and low-income populations. That is, one should perform a less rigorous analysis of proposals with a clearly small potential for impact, while devoting correspondingly more attention to actions where the potential for significant disruption is greater.

Attention should focus on identifying and evaluating impacts to minority and low-income populations that would be different from the impact on the general population. A qualitative assessment can often be sufficient to provide the decision maker with information on which to base informed decisions. Where such differences are trivial, include only enough discussion to show why further study is not warranted.

Approach any investigation of impacts on minority or low-income populations as a subset of impacts on the general population. If appropriate, this should be done on a resource-by-resource basis (e.g., air quality, water quality) or impact area (e.g., health impacts, facility, food sources, accidents, cultural disruption). Any special mechanisms by which an impact could affect a minority or low-income population might also need to be described. The size of the population and its geographic area should be indicated.

Consider, as appropriate, whether the proposal would:

- Affect or deny access to any natural resource on which the minority or low-income population (but not the general population) depends for cultural, religious, or economic reasons (e.g., a plant from which art is made, and perhaps sold for profit).
- Affect a minority or low-income population's food source, by reducing its abundance (e.g., development that would eliminate land habitat where game animals forage, or that would increase silt in a stream that is fished).

Unique pathways, exposures, and cultural practices   In assessing environmental impacts on minority and low-income populations, one should investigate the effects, based on considerations such as special pathways, exposures, and cultural practices. Table 4.4 presents factors useful in identifying unique pathways, exposures, or cultural practices that might need to be considered. Table 4.5 provides a definition of subsistence consumption cited in Table 4.4.

Considering cumulative impacts   As appropriate, agencies should consider the potential for *multiple* or *cumulative exposures*.[12] "Multiple exposure" and "cumulative exposure" are defined in Table 4.6.

In assessing cumulative impacts, practitioners need to recognize that minority and low-income populations might be affected by past, present, or reasonably foreseeable future actions in a manner different from that experienced by the general population.

Determining whether impacts are high and adverse   Agencies are expected to integrate their analyses of EJ concerns in an appropriate manner so as to be clear and concise within the general format of the EA/EIS outlined in the regulations (§1502.10 and §1508.9[b]).[13]

The analysis should clearly indicate, along with the basis for the conclusion, whether there are any significant impacts to a minority or low-income population. To this end, the analysis should indicate whether high and adverse impacts on a minority or low-income population would appreciably exceed the same type of impacts on the general population.

Determining whether impacts are disproportionate   Compare any high and adverse impacts on minority and low-income populations to

- The same type of impacts on the general population (e.g., air quality to air quality).
- Impacts on the general population, not on another subset of the general population (e.g., not on a non-minority or high-income population)

Table 4.4 Factors Useful in Identifying Unique Pathways, Exposures,
or Cultural Practices

Does the minority or low-income population (but not the general
population) use a natural resource or area for cultural, religious, or
economic reasons? Such uses could include
  • Plants for ceremonial or medicinal purposes, or from which art is made
    and perhaps sold for a profit
  • Plant-gathering or clay-procurement areas
  • Ceremonial site

Are exposure pathways or rates of exposure for minority and low-income
populations different from exposure pathways or rates for the general
population? Different pathways or rates could result from variations in
  • Physical location of a population's residences, workplaces, or schools
  • Dietary practices such as consumption of wild plants, or subsistence
    hunting, fishing, or farming
  • Differential selection of foods that might have high concentrations of
    contaminants (e.g., bottom-feeding fish or fish that feed on bottom-
    feeding organisms can bioconcentrate fat-soluble contaminants from
    sediments, and organ meats, such as elk liver, might have
    bioaccumulated such contaminants)
  • Water supplies, such as use of surface or well water for drinking or
    irrigation

Are there any known health, social, or economic conditions of a minority or
low-income population that would result in a greater impact? For
example, would there be a greater frequency of dose or greater impact
from a dose, over a pathway shared with the general population? Such
conditions could involve
  • Different access to public services such as paved roads (unpaved roads
    increase exposure to contaminated fugitive dust)
  • Different access to health care (e.g., poor control of asthma can increase
    susceptibility to particulate matter in air)

Bear in mind that a potentially beneficial impact on the general population may present a disproportionately high and adverse impact on a
minority and low-income population. For example, a highway might benefit people as a whole, yet so disrupt a minority or low-income population
as to constitute a disproportionately high and adverse impact.

Assessing significance and mitigation measures   On completing the
analysis, the agency needs to determine whether potential impacts on
minority and low-income populations are high and adverse, using the
criteria specified for assessing significance in the CEQ NEPA regulations
(§1508.27).

Table 4.5 Definitions of Subsistence Consumption

In its 1997 Environmental Justice Guidance under the National Environmental Policy Act, the CEQ issued guidance on key terms related to subsistence consumption as inserts in a reprinting of Executive Order 12898. The following guidance was developed by an Interagency Working Group on Environmental Justice, established by the executive order and chaired by the EPA:

- Subsistence consumption of fish and wildlife—Dependence by a minority population, low- income population, Indian tribe or subgroup of such populations on indigenous fish, vegetation, and/or wildlife, as the principal portion of their diet.
- Differential patterns of subsistence consumption—Differences in rates and/or patterns of subsistence consumption by minority populations, low-income populations, and Indian tribes as compared to rates and patterns of consumption of the general population.

Table 4.6 Definitions of Multiple and Cumulative Environmental Exposures

In its 1997 Environmental Justice Guidance under the National Environmental Policy Act, the CEQ presented definitions related to multiple and cumulative environmental exposure as inserts in a reprinting of Executive Order 12898. The following proposed definitions were developed by the Interagency Working Group on Environmental Justice, established by the Executive Order and chaired by the EPA:

- Multiple environmental exposure—Exposure to any combination of two or more chemical, biological, physical, or radiological agents (or two or more agents from two or more of these categories) from single or multiple sources that have the potential for deleterious effects to the environment and/or human health.
- Cumulative environmental exposure—Exposure to one or more chemical, biological, physical, or radiological agents across environmental media (e.g., air, water, soil) from single or multiple sources, over time in one or more locations, that have the potential for deleterious effects to the environment and/or human health.

The presidential memorandum accompanying Executive Order 12898 states that

> *Mitigation measures outlined or analyzed in an environmental assessment, environmental impact statement, or record of decision, whenever feasible, should address significant and adverse environmental effects of proposed federal actions on minority communities and low-income communities.*[14]

Mitigation measures include steps to avoid, minimize, rectify, reduce, or eliminate the adverse impact (§1508.20). The goal in mitigating disproportionately high and adverse effects is not to distribute the impacts proportionally or divert them to a non-minority or higher-income population. Instead, measures or alternatives should be developed to mitigate effects on both the general population and minority or low-income populations. In other words, the goal of mitigation is not to move the impacts around, but rather to identify practicable means to meet the purpose and need for taking action while avoiding or reducing undesirable environmental effects.[15]

Public participation

Presidential Executive Order 12898 discusses the importance of public participation in addressing environmental justice issues for proposed federal actions.[16]

Notifications

Both notices of EA preparation and notices of intent (NOIs) to prepare an EIS provide useful mechanisms for generating early participation in the NEPA process by minority and low-income populations. To be effective, however, the affected populations must receive the notices and understand their role in the NEPA process.

For EAs, it can be useful to notify not only the host and any potentially affected tribes and states of the agency's determination to prepare an EA, but also to notify potentially interested minority and low-income populations (and other stakeholders). When practical, explain the role of stakeholder participation in the EA process.

For EISs, consider disseminating NOIs not only through the *Federal Register* and major media outlets, but also through local distribution outlets.

Public meetings   To help ensure that meeting places and times are appropriate, it is advisable to check on meeting times and places of local community groups to minimize scheduling conflicts with any NEPA-related public meetings.

Agencies should consider scheduling public meetings to accommodate people who work night or weekend shifts. Churches or other places of worship, schools, community centers, public meeting halls, or local restaurants or hotels can be used as meeting places.

Agency officials might want to consider giving special consideration to segments of potentially affected populations that live in remote locations. For example, it might be beneficial to conduct a "road-show" style of public scoping meetings, traveling to several towns of various sizes in a short period to broaden the opportunities for participation among minority and low-income populations in small towns and rural areas.

Seeking comments   As appropriate, involve minority and low-income populations in identifying alternatives (including the environmentally preferable alternative) and issues for analysis that would address their special concerns. Minority and low-income populations may be the sole authoritative sources of information on their cultural characteristics.

Summary

Performed poorly, an SIA may be nothing more than a public relations exercise for unscrupulous developers. But when properly implemented, SIAs can provide a realistic appraisal of how a community and its citizenry are likely to be affected in the future. A well-investigated SIA can alleviate much of emotional uncertainty surrounding the project.

Like the NEPA and EIA, an SIA is not intended to hamper development. Instead, it can maximize the potential benefit for most or all of the involved parties. For the local community, it means reducing social impacts while increasing community benefits. For the developer, it can minimize social impacts and the costs of rectifying these effects in the future. In fact, an SIA can increase the legitimacy of the project and may actually facilitate the project development. It may also reduce impacts on the workforce, while increasing productivity and reducing disruption. Ultimately, the effectiveness of an SIA rests on the integrity of the SIA practitioner and the wiliness of the decision maker to seriously consider the analysis. Finally, all participants must understand that an SIA alone does not lead to a final discrete decision. Socioeconomic decisions are, in the final analysis, essentially political.

Endnotes

1. Dixon, M., *What Happened to Fairbanks: The Effects of the Trans-Alaska Oil Pipeline on the Community of Fairbanks, Alaska.* Boulder, CO: Westview Press, 1978.
2. Inter-organizational Committee on Guidelines and Principles for Social Impact Assessment, Guidelines and principles for social impact assessment, *Impact Assessment*, 12(2), 107–152, 1994, Belhaven, NC.
3. Burdge, R. J., and Vanclay, F., Social impact assessment: a contribution to the state of the art series, *Impact Assessment*, 14(1), 59–86, 1996.
4. Inter-organizational Committee on Guidelines and Principles for Social Impact Assessment, Guidelines and principles for social impact assessment, *Impact Assessment*, 12(2), 107–152, 1994, Belhaven, NC.

5. Presidential Executive Order 12898, Federal Actions to Address Environmental Justice in Minority Populations and Low-Income Populations, signed by President Clinton in February 1994; CEQ Guidance, Environmental Justice Guidance under the National Environmental Policy Act, December 1997; US Environmental Protection Agency, Guidance for Incorporating Environmental Justice Concerns in EPA's NEPA Compliance Analyses, 1998; EPA Guidance for Consideration of Environmental Justice in Clean Air Act Section 309 Reviews, 1999; US Department of Energy, Draft Guidance on Incorporating Environmental Justice Considerations into the Department of Energy's National Environmental Policy Act Process, April 2000.
6. US Environmental Protection Agency, Guidance for Incorporating Environmental Justice Concerns in EPA's NEPA Compliance Analyses, 1998; EPA Guidance for Consideration of Environmental Justice in Clean Air Act Section 309 Reviews, 1999.
7. Presidential Executive Order 12898, Federal Actions to Address Environmental Justice in Minority Populations and Low-Income Populations, February 1994.
8. CEQ (Council for Environmental Quality), Environmental Justice Guidance under the National Environmental Policy Act, December 1997.
9. US Environmental Protection Agency, Guidance for Incorporating Environmental Justice Concerns in EPA's NEPA Compliance Analyses, 1998.
10. US Environmental Protection Agency, EPA Guidance for Consideration of Environmental Justice in Clean Air Act Section 309 Reviews, 1999.
11. CEQ (Council for Environmental Quality), Environmental Justice Guidance under the National Environmental Policy Act, December 1997.
12. CEQ (Council for Environmental Quality), Environmental Justice Guidance under the National Environmental Policy Act, December 1997.
13. CEQ (Council for Environmental Quality), Environmental Justice Guidance under the National Environmental Policy Act, December 1997.
14. Presidential Executive Order 12898, Federal Actions to Address Environmental Justice in Minority Populations and Low-Income Populations, February 1994.
15. EPA, Guidance for Incorporating Environmental Justice Concerns in EPA's NEPA Compliance Analysis, 1998.
16. Presidential Executive Order 12898, Federal Actions to Address Environmental Justice in Minority Populations and Low-Income Populations, February 1994.
17. Eccleston, C. H. *NEPA and Environmental Planning: Tools, Techniques, and Approaches for Practitioners*, Boca Raton, FL: CRC Press, pp. 236–237, 2008.
18. Eccleston, C. H. *NEPA and Environmental Planning: Tools, Techniques, and Approaches for Practitioners*, Chapter 2, Boca Raton, FL: CRC Press, 2009.

chapter five

The international environmental impact assessment process

> *It ain't what you know that gets you in trouble. It's what you know for sure that just ain't so.*
>
> **—Mark Twain**

Paoletto has suggested a set of minimal elements that a typical environmental impact assessment (EIA) document should contain (Table 5.1).[1] Meanwhile, the International Association of Impact Assessment (IAIA) has gone beyond this fundamental guidance by producing a set of 14 *governing principles* for all EIA processes (Table 5.2). These principles apply to all stages of an EIA or strategic environmental assessment (SEA) process.

The IAIA has also identified ten *operating principles* applicable to all EIA processes (Table 5.3).[2] These operating principles describe how the basic principles outlined in Table 5.4 should be applied to the main steps and specific activities of the EIA process (i.e., screening, scoping, identification of impacts, and assessment of alternatives).

Comparison of NEPA with other EIA processes

A final decision to pursue an action generally cannot be made before the EIA process has been completed. This requirement aims to ensure consideration of environmental impacts during the decision-making process. In a few countries, a final decision can actually be made before the EIA process has been completed.

A federal agency generally formulates a proposal and is responsible for preparing the EIS. In practice, consultants are often used to assist the agency in preparing the EIS. Allowing the proposing agency to prepare the EIS has been criticized as potentially biasing the analysis. To promote a more objective analysis, some countries have their EIAs prepared by independent agencies.

A U.S. NEPA EIS is only required for *federal* actions that may significantly affect environmental quality; this restriction also includes private actions that are enabled by a federal agency (funded, authorized, approved by a federal agency). Additionally, the majority of U.S. states have their own state environmental policy acts (SEPAs), some of which have requirements

Table 5.1 Typical Steps of a Project-Specific EIA Process

Impact identification: Involves a broad analysis of the impacts of project activities with a view to identifying those that are worthy of a detailed study.

Baseline study: Involves collection of detailed information and data on the condition of the project area prior to the project's implementation.

Impact evaluation: Is performed whenever possible in quantitative terms and should include the working-out of potential mitigation measures.

Assessment: Assessing the environmental losses and gains, with economic costs and benefits for each analyzed alternative.

Documentation: A document detailing the EIA process and conclusions regarding the significance of potential impacts.

Decision making: The document is transmitted to the decision maker, who will accept one of the project alternatives, request further study, or reject the proposed action altogether.

Post audits: Determine how close to reality the EIA predictions were.

more rigorous than those of the NEPA. These SEPAs often encompass private actions that occur within those states. In contrast, European Union (EU) assessments can be triggered by many types of private or public actions that may significantly affect the environment; categories of actions requiring some level of EIA analysis have been established. Under both the US and EU processes, private applicants generally supply data for the analysis. Both the US and EU have well defined public participation requirements.

In the US, a programmatic EIS can be prepared for programs, policies, and plans. The EU countries do not recognize a programmatic EIA. Instead, they prepare strategic EIAs that are somewhat similar to programmatic EISs (See Section 5.5).

The NEPA implementing regulations establish basic requirements that each US agency must follow. Each agency is required to supplement these basic requirements with its own specific implementing regulations. Similarly, EU countries establish basic EIA requirements that each member state must follow. Each nation can supplement these basic requirements with its own requirements.

In the United States, the federal court system is used to litigate legal challenges involving NEPA compliance issues. In the European Union, the European Court of Justice decides legal challenges involving the EIA process.

Neither US nor EU impact assessment processes require that one choose an alternative that protects environmental quality. Both processes are largely founded on the premise that the analysis will lead to more informed decision making and, ultimately, to decisions that will protect the environment.

Table 5.2 Governing EIA Principles

Purpose	Supports informed decision making that results in appropriate levels of environmental protection and well-being of the community
Rigorous	Applies best practicable science, employing methodologies and techniques appropriate to address the problem under investigation
Systematic	Ensures full consideration of all relevant information on the affected environment, proposed alternatives and their impacts, and measures necessary to monitor and investigate residual effects
Interdisciplinary	Provides appropriate techniques and experts in the relevant biophysical and socioeconomic disciplines including the use of relevant traditional knowledge
Practical	Applies information and outputs that assist with problem solving, and that are both acceptable to proponents and can be implemented by the practitioners
Participative	Provides appropriate opportunities to inform and involve interested and affected publics, and ensure that their inputs and concerns are addressed explicitly in the documentation and decision-making process
Relevant	Provides sufficient, reliable, and usable information for development planning and decision making
Cost-effective	Implements a process that achieves the EIA objectives within the limits of available information, time, resources, and methodologies
Efficient	Imposes minimum cost burdens, in terms of time and finance on proponents and participants, that are consistent with meeting accepted requirements and EIA objectives
Focused	Concentrates on significant environmental effects and key issues (i.e., issues that need to be considered during the decision-making process)
Adaptive	Can be adjusted to the realities, issues, and circumstances of the proposals under review without compromising the integrity of the process; and is iterative, incorporating lessons learned throughout the proposal's life cycle
Credible	Can be carried out with professionalism, rigor, fairness, objectivity, impartiality, and balance, and be subject to independent checks and verification
Integrated	Addresses the interrelationships of social, economic, and biophysical aspects of the environment
Transparent	Presents clear, easily understood EIA requirements; ensures public access to information; identifies factors that are to be taken into account in decision making; and acknowledges limitations, problems, and difficulties

Source: International Association for Impact Assessment.

Table 5.3 IAIA Operating Principles Underlying EIA Process

Screening	Determines whether a proposal should be subject to an EIA and, if so, at what level of detail
Scoping	Identifies the issues and impacts that are likely to be important
Examination of alternatives	Establishes the preferred or most environmentally sound and benign option for achieving the proposal's objectives
Impact analysis	Identifies and predicts the likely environmental, social, and other related effects of the proposal
Mitigation and impact management	Establishes measures necessary to avoid, minimize, or offset predicted adverse impacts and, where appropriate, incorporate these into an environmental management plan or system
Evaluation of significance	Determines relative importance and acceptability of residual impacts (i.e., impacts that cannot be mitigated)
Preparation of environmental impact statement or report	Documents clearly and impartially the impacts of the proposal, the proposed measures for mitigation, the significance of effects, and the concerns of the interested public and the communities affected by the proposal
Review of the EIS	Determines whether the report meets its terms of reference; provides a satisfactory assessment of the proposal(s); and contains the information required for decision making
Decision making	Approves or rejects the proposal and establishes the terms and conditions for its implementation
Follow-up	Ensures that the terms and conditions of approval are met; monitors the impacts of development and the effectiveness of mitigation measures; strengthens future EIA applications and mitigation measures; and, as necessary, undertakes environmental audit and process evaluation to optimize environmental management. [*Note:* When monitoring, evaluation and management plan indicators are designed, it is desirable, whenever feasible, that they also contribute to local, national, and global monitoring of the state of the environment and to sustainable development.]

Source: International Association for Impact Assessment.

Table 5.4 Basic Elements Addressed in a Typical EIA

A brief nontechnical summary of the significant issues and findings
Any uncertainties and gaps in information
Description of the proposal
Description of the affected environment
Description of reasonable alternatives
Assessment of potential environmental impacts of the proposal (proposed action
  and alternatives), including short-term and long-term effects, and the direct,
  indirect, and cumulative impacts
Assessment of whether the environment of any other state/province or areas
  beyond national jurisdiction are likely to be affected by the proposal
Description of practical measures (including their effectiveness) for mitigating
  significant adverse environmental impacts of the proposed activity and alternatives

Of late, NEPA lawsuits are increasingly involving climate change issues. For proposals that could significantly affect the climate, US courts appear to be moving in the direction of requiring some degree of analysis. As applicable, EIAs in the European Union must address its emission trading system.

Comparison with World Bank

The NEPA recognizes only one instrument for investigating proposals that are deemed to result in a significant environmental impact—the EIS. The World Bank recognizes a number of instruments that depend on the nature of the problem; these instruments include: EIAs, regional/sector environment assessments, environmental audits, risk/hazard assessments, and environmental management plans.

In the United States, a notice of intent (in addition to other notices that an agency's NEPA implementing regulations require) must be published in the *Federal Register*. The public is kept abreast of the status and is afforded an opportunity to participate over the course of the EIS, beginning with the public scoping process.

The World Bank uses various mechanisms for publicizing its EIA process, including publishing a status of all projects, by country, on its Website. EIAs are circulated, normally for 120 days, to potentially affected parties in their local language. The bank directly engages people for their views and comments.

Programmatic and strategic Environmental assessments

A relatively recent innovation involves the concept of *programmatic environmental assessments (PEAs)* and *strategic environmental assessments (SEAs)*. The strategic environmental assessment can be defined as[3]

> *A process of anticipating and addressing the potential
> environmental consequences of proposed initiatives at
> higher levels of decision-making. It aims at integrating
> environmental considerations into the earliest phase of
> policy, plan or program development, on a par with eco-
> nomic and social considerations.*

SEAs can provide policymakers with a valuable tool in formulating policies and plans. In essence, an SEA extends the application of EIA to the level of policies, plans, and programs (PPPs). A key distinction between a project-level EIA and an SEA is that an SEA can be applied to PPPs at an earlier stage than individual projects. Thus, SEA allows for environmental considerations and objectives to be viewed proactively, as inherent elements of the planning process, rather than just as problems to be mitigated after other development decisions have been made.

Increasingly, SEAs are being used to shape the initial stages of decision making to assess the consequences of PPPs. Countries such as the United States, Australia, Canada, Denmark, and New Zealand already apply mechanisms similar to SEAs in developing plans and policies.[4] Some aid agencies in Africa have also started to use them.[5] SEAs are also being recognized as a pro-active tool for promoting sustainable development that may also serve to reduce the number of required project-specific EIAs. Planners can use it as a method to assess different ways for accomplishing sustainability policies (Chapter 3 provides an overview of the concept of sustainability).

Much of the world, including the European Union, does not formally recognize a *programmatic* EIA, which is a concept defined in the US NEPA regulations; the concept of the *strategic* EIA is recognized throughout the European Union and is gaining acceptance in many other countries. Conversely, the United States and many other nations do not formally recognize the concept of the *strategic* EIA. In practice, the two concepts are similar. While there are no universally accepted criteria for differentiating them, the two concepts are most commonly applied in the following manner.

A *programmatic* assessment tends to be prepared for proposals that involve the development of a definite policy, plan, or program. In contrast, a *strategic* assessment tends to denote a scope of analysis that is one level higher than that of a programmatic assessment. More to the point, a strategic assessment defines a high-level direction or strategy for a nation (e.g., national energy, agricultural, or water strategy). In theory, once this direction/strategy is defined, a *programmatic* assessment could then be prepared to consider a specific policy, plan, and in particular, a program for implementing the strategy defined in the SEA; the *programmatic* EIA can be tiered from the SEA. Once the *programmatic* EIA has defined a program-level course of action (such as a specific nuclear energy or solar program for implementing the energy strategy), more standard EIAs can then be tiered off the *programmatic* EIA to assess site-specific impacts of a project such as construction of a nuclear power plant or solar array farm.

Goals of the SEA

According to Sadler, an SEA should be prepared to

- Focus project-specific EIAs by ensuring that issues of need and alternatives are addressed at the appropriate policy, plan, or program level.
- Improve the scope and assessment of cumulative impacts, particularly where large projects stimulate secondary development and where many small developments not requiring EIAs may occur.
- Facilitate the application of sustainable principles and guidelines, for example, by focusing on the maintenance of a chosen level of environmental quality, rather than by minimizing individual impacts.

Performance criteria

As depicted in Table 5.5, the IAIA has established performance criteria that SEA analyses should meet.[6]

The relationship between EIA and SEA

The difference between EIA and SEA processes is evident in the scale of their frameworks. When compared to a project-specific EIA, the scope of an SEA tends to be broader, both temporally and geographically, and allows consideration of alternatives and a higher programmatic view of the bigger picture.

Ideally, a project-specific EIA should be prepared once a policy has been established via an SEA. The EIA provides information about the likely environmental impacts of an individual project and is useful in implementing mitigation measures. For example, if a government agency decides to develop a national wind power program, EIAs can be used to minimize the environmental damage from building specific power stations, but cannot practically address the more fundamental questions regarding the design of a national wind power program. In contrast, an SEA could effectively detail overall policy and investigate the programmatic impacts associated with such a policy, but is not an appropriate tool for evaluating site-specific impacts, alternatives, or mitigation measures. Figure 5.1 depicts the relationship of various levels of EIA assessments. [*Note:* The types of assessments shown in this figure are not recognized by many nations.]

Table 5.5 SEA Performance Criteria

Integrated	Ensures adequate environmental assessment of all strategic decisions relevant to the achievement of sustainable development.
	Addresses interrelationships of biophysical, social, and economic aspects.
	Tiers to policies in relevant sectors and (transboundary) regions and, where appropriate, to project EIA and decision making.
Sustainability-led	Facilitates identification of development options and alternative proposals that are more sustainable.
Focused	Provides sufficient, reliable and usable information for development planning and decision making.
	Concentrates on key issues of sustainable development.
	Customizes analysis to the characteristics of the decision-making process.
	Is cost- and time-effective.
Accountable	Is carried out with professionalism, rigor, fairness, impartiality, and balance.
	Is subject to independent checks and verification
	Documents and justifies how sustainability issues were taken into account in decision making.
Participative	Informs and involves interested and affected public and government bodies throughout the decision-making process.
	Addresses their inputs and concerns in documentation and decision making.
	Provides clear, easily understood information and requirements, and ensures sufficient access to all relevant information.
Iterative	Ensures availability of the assessment results early enough to influence the decision-making process and inspire future planning.
	Provides sufficient information on the actual impacts of implementing a strategic decision, to judge whether this decision should be amended and to provide a basis for future decisions.

Comparison of SEA and EIA

McDonald and Brown have written that "EIA tends to focus on the mitigation of impacts of proposed activities rather than determining their justification and siting."[7]

Perhaps the most significant area in which SEAs differ from EIAs is that an SEA is a pro-active tool for environmental management, whereas an EIA is used to assess specific development proposals. Some fundamental differences between SEA and EIAs are summarized in Table 5.6.

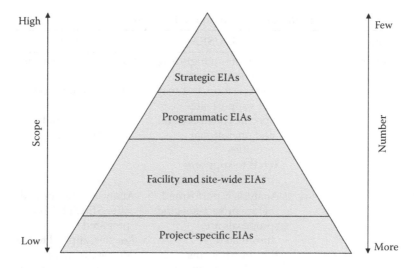

Figure 5.1 Relationship between different levels of EIAs.

SEA and EIA also tend to be applied at different stages of plans and policies and to different levels of decision making. Such a *tiered approach* is employed in New Zealand, the European Union, and the United States.[8] Under a tiered approach, SEAs are used to formulate strategies and policies in a proactive way. These policies and strategies create a framework against which specific development proposals and projects can then be assessed using the EIA.

Swedish planners have used the SEA to ensure that plans and environmental goals encourage sustainable development.[9]

Guidance for preparing programmatic NEPA assessments

This guidance is intended to assist practitioners in preparing U.S. NEPA *programmatic* environmental impact statements (EISs) and *programmatic* environmental assessments (EAs). Although this guidance focuses primarily on programmatic EISs, many U.S. agencies make effective use of programmatic EAs, and preparing a programmatic EA is appropriate when an agency needs to determine whether a broad proposed action or subsequent implementing actions require an EIS. As with any type of EA, agencies may prepare a programmatic EA to facilitate the preparation of a required EIS, to aid compliance with NEPA, or at any time to further the purposes of NEPA. While the following direction applies to US NEPA documents, much of this guidance is also applicable (with slight modifications or nuances) to international EIA processes.

Table 5.6 Comparison of EIA with SEA

Attribute	PEA/SEA	EIA
Decision making	Formulates a high-level direction; supports site selection, but may not support site-specific construction and operational activities Tends to supports multiple decisions	Supports detailed site selection decisions; supports site specific decisions such as construction and operational activities
Timing	Analysis is performed before project-specific proposals are formulated *Proactive*; informs stakeholders about future development proposals	Analysis is performed after SEA/PEA is prepared *Reactive*; applied to the development of a site-specific proposal
Spatial bounds	Considers large areas, regions, or national development; focuses on impacts of national or regional significance	Focuses on a specific project site; focuses on site-specific impacts
Temporal bounds	Longer shelf life Tends to be a continuing process over a life cycle aimed at providing information at the right time	Shorter shelf life Tends to have a defined beginning and endpoint
Level of uncertainty	Greater degree of uncertainty	Less uncertainty
Impact analysis	Evaluates broad assessment of cumulative impacts and identifies sustainable development issues	Tends to focus more on site-specific cumulative impacts, and direct and indirect impacts
Degree of quantitative analysis	Tends to be more descriptive and qualitative	Tends to be more quantitative
Focus of analysis	Focuses on maintaining a chosen level of environmental quality	Focuses on mitigating impacts

Table 5.6 Comparison of EIA with SEA (Continued)

Attribute	PEA/SEA	EIA
Level of detail	Has a broader perspective and a correspondingly lower level of detail, providing an overall vision	Has a narrow (project-specific) perspective with a higher level of detail
Mitigation	Considers general mitigation measures	Considers more specific mitigation measures
Interim actions	Frequently involves interim actions (actions related to the EIA that need to occur while the analysis is being prepared)	Less likely to involve interim actions Interim actions (actions that need to be taken before a final decision is made) are often prohibited or are severely restricted

Programmatic analyses and tiering

A *programmatic* analysis is a broadly scoped technique that assesses the environmental impacts of federal actions across a span of conditions, such as facilities, geographic regions, or multi-project programs. While the US Council on Environmental Quality's (CEQ) NEPA regulations[10] do not define the term "programmatic analyses," the regulations do address analyses of broad actions and they address tiering, the linkage between the broad action and subsequent, more focused or specific proposed actions. For example, the regulations state

> *Environmental impact statements may be prepared, and are sometimes required, for broad Federal actions such as the adoption of new agency programs or regulations. Agencies shall prepare statements on broad actions so that they are relevant to policy and are timed to coincide with meaningful points in agency planning and decisionmaking.*[11]

The regulations also identify three ways to evaluate proposals that agencies may find useful when preparing statements on broad actions:[12]

1. *Geographically,* including actions occurring in the same general location, such as a body of water, region, or metropolitan area.

2. *Generically,* including actions that have relevant similarities, such as common timing, impacts, alternatives, methods of implementation, media, or subject matter
3. *By stage of technological development,* including federal or federally assisted research, development, or demonstration programs for new technologies that, if implemented, could significantly affect the quality of the human environment

Where a broad policy, plan, program, or project will later be translated into site-specific projects, subsequent investigations are appropriate and are referred to as tiered analyses. Tiering is a method "to relate broad and narrow actions and to avoid duplication and delay." It is described in the NEPA regulations:

> *Agencies are encouraged to tier their environmental impact statements to eliminate repetitive discussions of the same issues and to focus on the actual issues ripe for decision at each level of environmental review (Sec. 1508.28). Whenever a broad environmental impact statement has been prepared (such as a program or policy statement) and a subsequent statement or environmental assessment is then prepared on an action included within the entire program or policy (such as a site specific action) the subsequent statement or environmental assessment need only summarize the issues discussed in the broader statement and incorporate discussions from the broader statement by reference and shall concentrate on the issues specific to the subsequent action. The subsequent document shall state where the earlier document is available. Tiering may also be appropriate for different stages of actions.[13]*

The NEPA regulations define tiering as[14]

> *... the coverage of general matters in broader environmental impact statements (such as national program or policy statements) with subsequent narrower statements or environmental analyses (such as regional or basin-wide program statements or ultimately site-specific statements) incorporating by reference the general discussions and concentrating solely on the issues specific to the statement subsequently prepared. Tiering is appropriate when the sequences of statements or analyses are:*

> (a) *From a program, plan, or policy environmental impact statement to a program, plan, or policy statement or analysis of lesser scope or to a site-specific statement or analysis.*
>
> (b) *From an environmental impact statement on a specific action at an early stage (such as need and site selection) to a supplement (which is preferred) or a subsequent statement or analysis at a later stage (such as environmental mitigation). Tiering in such cases is appropriate when it helps the lead agency to focus on the issues which are ripe for decision and exclude from consideration issues already decided or not yet ripe.*

The similarities and differences between programmatic assessments and tiering

The concepts of programmatic and tiered assessments differ in both their focus and scope. Table 5.7 depicts some of the basic differences between programmatic and subsequent, often project-level, tiered analyses.

Addressing deferred issues

The programmatic NEPA analysis and documentation forms the basis for tiering future NEPA documents focusing on specific facets under review.[15] Also important to consider is when, not whether, the detailed evaluation of a tiered project should take place. In *California v. Block,* the 9th Circuit Court considered this and stated that "[t]he critical inquiry in considering the adequacy of an EIS prepared for a large scale, multi-step project is not that the project's site-specific impact should be evaluated in detail, but when such detailed evaluation should occur." The court went on to decide that "[t]his threshold is reached when, as a practical matter, the agency proposes to make an "irreversible and irretrievable commitment of the availability of resources" to a project at a particular site.[16]

The relationship between the programmatic document and future or subsequent tiered documents should be described in the programmatic document (e.g., programmatic EA or EIS). Decisions or analyses that are deferred to future documents should be articulated in the programmatic document. Agencies should ensure that "thresholds" or "triggers" that will be considered in determining when the tiered analysis will be conducted are clear to all interested parties.

Uses of programmatic analyses

Project-specific or site-specific NEPA analyses can be aided by first completing a programmatic NEPA analysis and document. A well-crafted

Table 5.7 Basic Differences Between Programmatic and Subsequent (Often Project-Level) Tiered Analyses

Consideration	Programmatic level	Subsequent or site-specific tiered level
Scope of decision	Broad, strategic program, policy, or plan	Detailed, site-specific, action oriented
Scope of action	Policy, program, planning	Projects
Decision timing	Early	Subsequent to programmatic decision
Nature of action	Strategic, conceptual	Construction, operations, site-specific actions
Alternatives	Broad, general, research, technologies, fiscal measures, economic, social, regulatory	Specific alternative locations, design, construction, operation, permits, site specific
Scope of impacts	Broad in scale and magnitude	Localized and specific
Scale of impacts	Macroscopic (e.g., at a national, regional, or landscape level)	Project level, mainly local
Time scale	Long- to medium-term (regulatory)	Medium- to short-term (permit)
Assessment of impacts	Qualitative and may be quantitative to the degree possible	Quantifiable
Key data sources	Existing national or regional statistical and trend data, policy and planning instruments	Field work, sample analysis, statistical data, local monitoring data

programmatic analysis can provide the basis for early or initial decisions, such as identifying geographically bounded areas within which future activities will be authorized so that the analysis of those future proposed activities can be limited to those areas. Another example is the analysis of cumulative impacts at a programmatic level that the subsequent, more specific, analysis could tier from, thus avoiding the need to reanalyze those cumulative impacts in each subsequent NEPA.

Programmatic NEPA analyses may also be useful to support policy- and planning-level decisions when there are limitations in available information and uncertainty regarding the timing, location, and environmental impacts of subsequent implementing action(s). That is, although agencies may not be able to predict with certainty the environmental consequences of specific program implementing actions, they may be able to make broad program decisions based on a programmatic analysis, provided the analysis adequately examines the reasonably foreseeable consequences of the proposed program.

One programmatic approach is to rely on the programmatic document to analyze the impacts and alternatives of the broad policy, plan, program, or project, and then use subsequent or tiered analyses to adequately assess the site-specific proposals based on the broader policy, plan, program, or project.

Some agencies may prepare a single NEPA document that is both programmatic and project specific in nature. Such a document is appropriate when an agency plans to make a broad program decision as well as decisions to implement one or more specific projects under the program. For example, the programmatic approach may address both the broad impacts of the proposed broad federal action and provide environmental analyses for subsequent decisions, such as determining the locations and designs of specific projects to implement a broad federal action. In such cases, as with any NEPA document, it is essential to clearly state the decisions the agency proposes to make based on the document and to ensure that the programmatic EIS contains appropriate analysis of impacts and alternatives to meet the adequacy test for both the program- and the site-specific proposal.

The programmatic approach may limit the depth of analyses to the broad impacts of a broad proposal, for example, the potential nation- or region-wide impacts of a new proposed federal policy, plan, program, or project. This approach is appropriate when the agency uses the programmatic analysis to support a decision on establishing the policy, plan, program, or project and will use future tiered decisions to implement specific projects under the broad policy, program, plan, or project. Such an approach is also appropriate in cases where a programmatic NEPA analysis and document is prepared for a specific action at an early stage with the expectation that subsequent, more specific analysis will be provided in subsequent NEPA documents that supplement or are tiered to the programmatic document.[17]

Table 5.8 provides some examples of various types of programmatic assessments.

Various types of programmatic analyses

The types of actions that programmatic NEPA analyses support fall into four major categories (Table 5.9).

For proposed actions falling within any of the categories shown in Table 5.9, agencies can use a phased decision-making strategy. That is,

Table 5.8 Examples of Various Types of Programmatic Actions and How They Are Designed to Support Subsequent Decisions

Example of broad or programmatic analysis	Why analysis was used	Trigger for further analysis or action
Geographic or regional agency policy (Example: New transportation system)	The programmatic (Tier I) EIS examines broad issues and various alternative modes of transportation (e.g., Tier I EIS for a high-speed rail system). It programmatically compares general locations and mode of transportation. It programmatically assesses general air-quality issues and general land use implications of the programmatic alternatives.	As site-specific projects are identified, each project will have a separate (Tier II) EA/EIS. The programmatic (Tier I) EIS specifies decisions that must be resolved in Tier II NEPA documents.
Agency policymaking (Example: Development of a policy for control of non-native invasive species throughout the federal park system)	The programmatic EIS evaluates broad policy issues such as potential locations, control strategies, programmatic mitigation measures, and large-scale cumulative impacts (e.g., a programmatic EIS establishes a policy for controlling non-native invasive vegetation and then prepares site-specific EISs/EAs for individual control projects). Other issues include avoiding segmentation of the analysis and tiered assessments.	The detection of new or spreading non-native, invasive species that is sufficient to trigger specific control programs requiring preparation of tiered site-specific EISs or environmental assessments (EA).

Table 5.8 Examples of Various Types of Programmatic Actions and How They
Are Designed to Support Subsequent Decisions (Continued)

Example of broad or programmatic analysis	Why analysis was used	Trigger for further analysis or action
Establishment of a program with a combination of known conditions (Example: A programmatic EA to evaluate a series of five routine payloads on expendable launch vehicles)	The programmatic EA analyzes common launch vehicles, multiple launch sites, and broad classes of payload risk, including plutonium-fueled payloads.	Each new launch is reviewed against the programmatic EA (launch vehicle, launch site, payload, and risk). Any actions or impacts that potentially fall outside those evaluated in the programmatic EA would require preparation of a project-specific EA to verify that they are covered. If the impacts are bounded within the scope of the programmatic EA, the site-specific EA is tiered off the programmatic EA. If the impacts are found to be potentially significant, the EA would lead to a decision to prepare an EIS.

agencies can prepare broad programmatic NEPA analyses from which they tier subsequent, more detailed, project-specific analyses for specific proposals implementing the program. Tiering avoids duplicative paperwork through the incorporation by reference of the information, discussions, and analyses from a broad NEPA analysis and document into the more specific one.

Direction from CEQ's forty questions

The "Forty Most Asked Questions Concerning CEQ's National Environmental Policy Act Regulations" provide the following direction for determining when EISs should be prepared for policies, plans, or programs:[18]

> *An EIS must be prepared if an agency proposes to imple-ment a specific policy, to adopt a plan for a group of*

related actions, or to implement a specific statutory pro-
gram or executive directive. Section 1508.18. In addition,
the adoption of official policy in the form of rules, regula-
tions and interpretations pursuant to the Administrative
Procedure Act, treaties, conventions, or other formal doc-
uments establishing governmental or agency policy which
will substantially alter agency programs, could require an
EIS.[19] *In all cases, the policy, plan, or program must have*
the potential for significantly affecting the quality of the
human environment in order to require an EIS. It should
be noted that a proposal 'may exist in fact as well as by
agency declaration that one exists.'[20]

The "Forty Most Asked Questions Concerning CEQ's National
Environmental Policy Act Regulations" provides further direction regarding when an area-wide or overview EIS is appropriate:[21]

Table 5.9 Four Major Categories of Programmatic NEPA Analyses

---

Adopting official policy: Decision to adopt official policy in a formal document
establishing an agency's policies that will result in or substantially alter agency
programs. Such a programmatic analysis should include a road map for future
agency actions with defined objectives, priorities, rules, and mechanisms to
implement objectives. Examples include
- National-level rulemaking
- Adoption of an agency-wide policy

Adopting formal plan: Decision to adopting formal plans, such as documents
that guide or prescribe alternative uses of federal resources, upon which future
agency actions will be based. For example, setting priorities, options, and
measures for future resource allocation according to resource suitability and
availability. Examples include
- Strategic planning linked to agency resource allocation
- Adoption of an agency plan for a group of related projects

Adopting agency program: Decision to proceed with a group of concerted
actions to implement a specific policy or plan; organized agenda with defined
objectives to be achieved during program implementation, with specification
of activities. Examples include
- A new agency mission or initiative
- Proposals to substantially redesign existing programs

Approving site-wide or area-wide actions: Decision to proceed with multiple
projects that are temporally or spatially connected and that will have a series
of associated subsequent or concurrent decisions. Examples include
- Several similar actions or projects in a region or nationwide
- A suite of ongoing, proposed, or reasonably foreseeable actions that share a
  common geography or timing, such as multiple functions within the
  boundaries of a large federal facility

> *The preparation of an area-wide or overview EIS may be particularly useful when similar actions, viewed with other reasonably foreseeable or proposed agency actions, share common timing or geography. For example, when a variety of energy projects may be located in a single watershed, or when a series of new energy technologies may be developed through federal funding, the overview or area-wide EIS would serve as a valuable and necessary analysis of the affected environment and the potential cumulative impacts of the reasonably foreseeable actions under that program or within that geographical area.*

Further guidance on preparing a programmatic NEPA analysis

Setting the scope of the programmatic document and any subsequent tiered document is necessary to ensure that the environmental analysis is appropriate for the decisions made. Thus, the scope of the actions, alternatives, and impacts requires careful and critical thinking and planning. The questions presented in Table 5.10 may be helpful when considering the need for, scope of, and approach to a programmatic NEPA process.

Table 5.10 Helpful Questions in Considering Need for, Scope of, and Approach to a Programmatic NEPA Process

What are the appropriate geographic and time scales for this analysis at the programmatic level to serve subsequent decisions?

What federal proposals and decisions need to be analyzed and made now and in the future regarding the broad federal action being proposed?

Will the decision based on the analysis be the final agency decision (e.g., develop a nationwide fuel mileage requirement) without subsequent decisions to establish the proposed regulation requiring subsequent NEPA analyses and documents?

How long will the analysis and decision be used? Consider how long the programmatic decision will be maintained and used for tiering subsequent actions and determine what factors may result in the need to supplement the analysis.

Is it necessary to analyze particular effects at a broader scale to facilitate analysis and/or decision making at the subsequent level, and is a programmatic NEPA the best way to do this? What are the meaningful decision points from proposal through implementation, and where are the most effective decision and NEPA analysis and document points in that continuum to address the potential for effects?

The scoping process provides a means to develop the scope of the analysis and assist the agency in addressing the questions presented in Table 5.10. Also note that the NEPA regulations require agencies to consider the following in determining the scope of the NEPA analysis:[22]

- Three types of actions: connected, cumulative, and similar
- Three types of alternatives: no action, other reasonable courses of actions, and mitigation
- Three types of impacts: direct, indirect, and cumulative

Analysis of actions

Broad federal actions are typically implemented throughout large geographic areas and are broadly focused over a long time frame. Connected and cumulative actions should be included in a programmatic NEPA analysis and document, and the responsible official must consider whether to also include any similar actions.[23] Although this consideration is the same as for project-specific proposals, the likely connected, cumulative, and similar actions for programmatic NEPA analyses and documents are also expected to be broad and programmatic in nature.

Analysis of alternatives

Alternatives are expected to reflect the level of the broad federal action being proposed. Programmatic NEPA documents should provide a well-defined scope for the next level of decision making. By clearly articulating the nature of subsequent tiered decisions, agencies can effectively craft the purpose and need, proposed action, and alternatives to a programmatic document. This allows agencies to develop focused alternatives in the programmatic document that limit the scope and alternative development of the subsequent tiered NEPA document.

The scope of the alternatives for a programmatic analysis includes no action, other reasonable courses of action, and mitigation. In situations where there is an existing program, plan, or policy, the no-action alternative will be the continuation of the present course of action until a new program, plan, or policy is developed.

When developing a programmatic policy, plan, program, or project NEPA analysis and document, alternatives can be considered and eliminated from detailed study. A brief written discussion of the reasons the alternatives were eliminated should be provided.

The examination of alternatives need not be exhaustive. As one court states, "a programmatic analysis would not require consideration of detailed alternatives with respect to each aspect of the plan —otherwise a programmatic analysis would be impossible to prepare and would merely be a vast series of site specific analyses."[24]

Impact assessment

Because programmatic analyses typically concern environmental effects over a broad geographic and time horizon, the depth and detail of impact analysis are expected to be broad and general for programmatic NEPA analyses and documents. The effects analyses will focus on the major impacts that might result from implementing a broad federal action, and especially on those resources or factors that are adversely impacted.

A subsequent site-specific EIS/EA will address site- or project-specific considerations.[25] The programmatic analysis looks to the environmental consequences of a project as a whole, and does not necessarily contain the same level of detail or specificity as a site- or project-specific EIS. A recent court case sheds some light on the degree to which impacts should be considered. The court stated that "an EIS for a programmatic plan must provide sufficient detail to foster informed decision-making, but that site-specific impacts need not be fully evaluated until a critical decision has been made to act on site development."[26]

In programmatic EISs, impacts will often be described as a range rather than as a specific amount, and will be discussed in a broad geographic and temporal context with particular emphasis on cumulative impacts. The scope and range of impacts may be more qualitative in nature than those found in project-specific NEPA documentation. While the agency's obligation remains to conduct a meaningful impact analysis in accordance with Section 102 (2)(C) of NEPA, the analysis should be commensurate with the decision to be made.

Addressing new information

The CEQ NEPA regulations specify that implementation of the action should be accompanied by monitoring in important cases and provide a procedural framework for keeping environmental analyses current by requiring that agencies prepare supplements if significant new information of relevance to the proposed action or its impacts is discovered.[27] In one case, the court declared that a "federal agency has a continuing duty to gather and evaluate new information relevant to the environmental impact of its actions, even after release of an environmental impact statement."[28]

When new information reaches an agency, it should be initially screened with respect to the following considerations:

- Does the new information pertain to a programmatic NEPA document that was prepared for a now completed decision-making process? Phrased in the alternative, are there any more decisions to be made by the agency that would use the original NEPA document to

meet all or a portion of the agency's NEPA compliance responsibilities for any upcoming decision?
- If the new information is relevant to a future decision for which the agency intends to rely upon the original programmatic analysis, then a second consideration must be addressed.
- The new information must be reviewed in order to determine whether it has any potential effect on the content of the original programmatic document, in terms of (1) the accuracy of the previously analyzed impacts (direct, indirect, or cumulative), or (2) the feasibility of the alternatives presented or their comparative analysis.

The agency is responsible for initially making a reasoned determination of whether the new information either raises significant new circumstances or information regarding environmental impacts, or involves substantial changes in the actions decided upon in the programmatic analysis, thereby requiring supplementation.[29]

If the original programmatic NEPA document was an EA, a similar reasoned determination is required. The focus of this determination, however, must show whether an EA and its finding of no significant impact (FONSI) still suffice, or whether an EIS is now necessary. The programmatic EA should be supplemented to address the new information and the FONSI reconsidered based on the agency's significant impact criteria, which would include consideration of the context and intensity of the effects of the programmatic action.

Interim actions

Agencies may be reluctant to conduct programmatic NEPA analysis because of the risk of delaying new or ongoing actions. This question arises more frequently for actions that fall within the scope of a programmatic EIS. The NEPA regulations provide a mechanism referred to as an *interim action* that may proceed while a programmatic assessment is underway, provided that certain criteria are met.[30] In general, proposed actions of relatively limited scope or scale that would have only local utility can normally be pursued as an interim action while the programmatic analysis is underway.

While the NEPA regulations address criteria for interim actions (cases where a proposed action needs to proceed while a programmatic EIS is underway), the regulations specifically state that[31]

> *While work on a required program environmental impact statement is in progress and the action is not covered by an existing program statement, agencies shall not undertake in the interim any major Federal action covered by*

> the program which may significantly affect the quality of
> the human environment unless such action:
> 1. Is justified independently of the program;
> 2. Is itself accompanied by an adequate environmental
>   impact statement; and
> 3. Will not prejudice the ultimate decision on the pro-
>   gram. Interim action prejudices the ultimate deci-
>   sion on the program when it tends to determine
>   subsequent development or limit alternatives.

As described in the first criterion (independent justification), agencies can take an interim action that could be undertaken irrespective of whether or how the program goes forward (assuming the other two criteria are met). For proposed interim actions involving changes to an existing facility within the scope of a programmatic EIS in preparation, the agency must establish that such an action is needed to allow the facility to fulfill its existing mission before decisions can be made and implemented using the programmatic EIS.

Under the second criterion, an EIS must be prepared for a proposed interim action that has the potential for significant environmental impact. In cases that do not involve significant impacts, an EA would be sufficient to provide adequate NEPA support for the decision and meet this second criterion.

As described in the third criterion (requiring the decision to be nonprejudicial to the programmatic decision), agencies may take an interim action when they determine that the proposed interim action would not tend to determine subsequent programmatic development, or limit programmatic alternatives. Furthermore, an agency does not need to suspend operations only because it has elected to prepare a programmatic NEPA document. For instance, in the case of an area-wide programmatic EIS, ongoing operations within the area may continue, and such operations would be considered under the no-action alternative in the programmatic EIS. To ensure that the programmatic action is not impermissibly segmented, before proceeding with such an interim action, agencies should consider the first and third criteria, and determine that the interim action is independently justified and will not prejudice the ultimate decision on the program.

Problems and limitations

In practice, application of the EIA process tends to focus on project-level planning (although it has also been infrequently applied to programs and strategic planning). A project in the context of EIA is "an individual development or other scheme as distinct from a suite of schemes or a strategy for development of a particular type or in a particular region."[32] When the EIA process is applied to broader programs or regional planning, it is often done through the related analytical process of strategic environmental assessment.

While an EIA process does not necessarily prevent a project from having an impact on the environment, it frequently does manage to minimize the severity of its adverse impacts.[33] Nonetheless, there are some fundamental problems with most processes. For example, some EIA processes only address alternatives to the proposed project in a *limited* manner; that is to say, by the project assessment stage, a number of options having potentially different environmental consequences from the chosen one are likely to already have been eliminated.

Optimism bias and the planning fallacy

Optimism bias is a demonstrated and systematic tendency for people to be overly optimistic about the outcome of an action. This includes overestimating the likelihood of positive events and underestimating the likelihood of negative ones. It is one of several kinds of positive illusions to which people are generally susceptible.

Planning fallacy

The *planning fallacy* is the tendency to underestimate the projected time required to complete a task. In one study, 37 college students were asked to estimate the completion time for their senior theses. The average estimate was 33.9 days. But only about 30% of the students completed their theses in the predicted amount of time; the average actual completion time was 55.5 days.[34]

Lovallo and Kahneman expanded this concept from the tendency to underestimate task-completion times to the more general tendency to underestimate the time, costs, and risks of future actions and at the same time overestimate the benefits of the same actions.[35] According to their definition, the planning fallacy results not only in schedule overruns, but also cost overruns and benefits that do not reach original expectations. Popular examples include construction of the Sydney Opera House and Boston's Big Dig project, both of which ran many years past their original schedules.

Flyvbjerg argued that what appears to be optimism bias may, on closer examination, be strategic misrepresentation.[36] He believes that planners frequently deliberately underestimate costs and overestimate benefits in order to get their projects approved, especially when projects are large and when organizational and political pressures are high. Kahneman and Lovallo maintained that optimism bias is the main problem.[37] Many studies have shown that this bias can cause people to take imprudent or even unacceptable risks.[38] This effect, however, is not universal; for instance, some people tend to overestimate the risks of low-frequency events, particularly negative ones.

Disadvantages of project-specific EIAs

Sadler describes some disadvantages of project-specific EIAs:[39]

- Restricted opportunities for effective public participation in planning or decision-making processes
- Restricted ability to address cumulative impacts, particularly for large development projects where secondary development could occur
- Limits an analysis in a stand-alone process, which may be poorly related to the project cycle

Because a project-level EIA often precludes consideration of alternative strategies, locations, and designs, at least one EIA practitioner argues that, in effect, "An EIA at the project level is essentially damage control."[40] Application of EIA at a more strategic level can promote a more effective assessment of alternatives and cumulative impacts at an earlier stage in the decision-making process. It can also facilitate consideration of a wider range of actions over a greater area.[41]

Endnotes

1. Paoletto, G. Lecture Notes Environmental Impact Assessments (EIA), http://www.gdrc.org/uem/eia/lecture-notes.html.
2. International Association for Impact Assessment, Principles of Environmental Impact Assessment Best Practice, http://iaia.org/Members/Publications/Guidelines_Principles/Principles%20of%20IA.PDF.
3. Sadler, B. and Verheem, R. *Strategic Environmental Assessment: Status, Challenge and Future Directions.* The Netherlands: Ministry of Housing, Spatial Planning and the Environment, 1996.
4. Sadler, B. and Verheem, R. *Strategic Environmental Assessment: Status, Challenge and Future Directions.* The Netherlands: Ministry of Housing, Spatial Planning and the Environment, 1996.
5. Goodland, R., Mercier, J.-R., and Muntemba, S., Eds.), Environmental Assessment (EA) in Africa: A World Bank Commitment, *Proceedings of the Durban (South Africa) Workshop,* June 25, 1995. Washington, DC: The World Bank, 1996.
6. International Association for Impact Assessment, *Principles of Environmental Impact Assessment Best Practice Best Practice,* http://iaia.org/Members/Publications/Guidelines_Principles/Principles%20of%20IA.PDF.
7. McDonald, G. T., and Brown, L., Going beyond environmental impact assessment: environmental input to planning and design. *Environmental Impact Assessment Review,* 15, 483–495, 1995.

8. CSIR (Council for Scientific and Industrial Research), CSIR Report, Strategic Environmental Assessment (SEA): A Primer, ENV/S-RR 96001, September 2, 1996.
9. Eggiman, B., Fysisk planering med strategisk miljöbedömning (SMB) för hållbarhet. En teoretisk diskussion och förslag till SMB-process med Stockholms stad sommodell. Karlskrona and Stockholm: Swedish Board of Housing, Building and Planning and Swedish Environmental Protection Agency, 2000.
10. 40 CFR §1500–1508.
11. 40 CFR §1502.4(b).
12. 40 CFR §1502.4(c).
13. 40 CFR §1508.28.
14. 40 CFR §1508.28.
15. *Greenpeace v. National Marine Fisheries Service*, 55 F. Supp. 2d 1248, 1273 (D. Wash. 1999).
16. *California v. Block*, 690 F.2d 753, 761, (9th Cir. 1982).
17. 40 CFR §1508.28(b).
18. 46 Fed. Reg. 18,026 (1981): Forty Most Asked Questions Concerning CEQ's National Environmental Policy Act Regulations, Question 24a. Environmental Impact Statements on Policies, Plans or Programs.
19. 40 CFR §1508.18.
20. 40 CFR §1508.23.
21. 46 Fed. Reg. 18,026 (1981): Forty Most Asked Questions Concerning CEQ's National Environmental Policy Act Regulations, Question 24b. When is an area-wide or overview EIS appropriate?
22. 40 CFR §1508.25.
23. 40 CFR §1508.25(a).
24. *Greenpeace v. National Marine Fisheries Service*, 55 F. Supp. 2d 1248, 1276 (D. Wash. 1999)
25. *Nevada v. Department of Energy*, 372 U.S. App. D.C. 432 (D.C. Cir. 2006).
26. *Citizens for Better Forestry v. United States Department of Agriculture*, 481 F. Supp. 2d 1059, 1086, (D. Cal. 2007).
27. 40 CFR §§1505.3 (monitoring); §1502.9 (supplementation).
28. *Seattle Audubon Society v. Moseley*, 798 F. Supp. 1473, (D. Wash. 1992).
29. 40 CFR §1502.9.
30. 40 CFR §1506.1.
31. 40 CFR §1506.1(a) and (c)
32. Therivel, R., Wilson, E., Thompson, S., Heaney, D., and Pritchard, D., *Strategic Environmental Assessment*, London: Earthscan, 1992.
33. Therivel, R., Wilson, E., Thompson, S., Heaney, D., and Pritchard, D., *Strategic Environmental Assessment*, London: Earthscan, 1992.
34. http://en.wikipedia.org/wiki/Planning_fallacy (accessed January 10, 2008).
35. Lovallo, D. and Kahneman, D., Delusions of success: How optimism undermines executives' decisions, *Harvard Business Review,* July issue, 56–63, 2003.
36. A running debate between Kahneman, D., Lovallo, D., and Flyvbjerg, B., in the *Harvard Business Review*, 2003.
37. Lovallo, D. and Kahneman, D., Delusions of success: How optimism undermines executives' decisions, *Harvard Business Review,* July issue, 56–63, 2003.

38. Armor, D.A. and Taylor, S.E., When predictions fail: The dilemma of unrealistic optimism. In Gilovich, Thomas (2002). *Heuristics and Biases: The Psychology of Intuitive Judgment*, Cambridge, UK: Cambridge University Press, 2002.
39. Sadler, B., Towards the Improved Effectiveness of Environmental Assessment, Executive Summary of Interim Report Prepared for IAIA'95, Durban, South Africa, 1995.
40. Symthe, R.B., Ph.D. personal communications, 2007.
41. Therivel, R., Wilson, E., Thompson, S., Heaney, D., and Pritchard, D., *Strategic Environmental Assessment*, London: Earthscan, 1992.

chapter six

Environmental management systems

A man generally has two reasons for doing a thing: one that sounds good, and a real one.

—J. P. Morgan

This chapter investigates the environmental management systems (EMSs) and how they can be used to implement the environmental assessments described in this book. We begin with an introduction to the ISO 14000 standards.

The ISO 14000 standards

The International Organization for Standardization, also referred to as ISO, is composed of representatives from nations around the world whose mission is to develop common standards for products and services. It is important to note that "ISO" is not an acronym. It is, in fact, derived from the Greek word *iso*, meaning "equal" or in this case, "equivalent standards." This section describes the ISO 14000 and ISO 14001 EMS.

The difference between ISO 14000 and ISO 14001

"ISO 14000" and "ISO 14001" are commonly cited in the literature. Not surprisingly, this has led to significant confusion concerning the differences between these two terms. As illustrated in Table 6.1, ISO 14000 refers to a series or a family of related environmental management standards (e.g., environmental management system, environmental labeling, environmental auditing, and environmental assessment of organizations). In contrast, ISO 14001 (the first element in the ISO 14000 series) deals exclusively with the requirements for establishing an ISO 14001-compliant EMS.

Improving the EMS system versus improving environmental performance

With respect to an ISO 14001 EMS, "environmental performance" typically refers to the act of reducing an environmental impact, such that the

229

Table 6.1 The ISO 14000 Series of Standards

Series	Explanation
ISO 14001	Requirements and guidance for using environmental management systems (EMSs).
ISO 14004	EMS general guidelines
ISO 14015	Assessment of organizations and sites
ISO 14020	Environmental labels and declarations
ISO 14031	Environmental performance evaluation guidelines
ISO 14040	Life-cycle assessment, pre-production planning, and environmental goal setting
ISO 14050	Definitions
ISO 14062	Improvements to environmental impact goals
ISO 14063	Environmental communications guidelines and examples

environmental quality is improved. Instead, ISO 14001 focuses on improving the process, which manages or administers an organizations functions and activities that can affect the environment. ISO 14001 does not actually require that an EMS improve environmental performance (i.e., reduce environmental impacts). For instance, ISO 14001 does not prescribe a particular level of pollution or environmental performance, require the use of particular technologies, or establish regulatory standards for environmental outcomes. In fact, some organizations engaged in similar activities may have widely different effects on the environment and yet all comply with ISO 14001.

The focus of an ISO 14001 EMS is on improving management processes, practices, and procedures that control an organization's functions and activities, which can affect the environment. The overarching intent is that by implementing a *management process* that administers an organization's functions, products, and services, and by continually improving this management system, this process will eventually lead to improved environmental performance. While this is generally true of organizations that are truly committed to the goal of improving environmental quality, it may not be true of an organization that lacks a serious commitment; in that case, an EMS may amount to nothing more than window dressing: to improve its image with the public and consumers.

It is important to note that adherence to the ISO 14000 standards does not alone release an organization from full compliance with other local or

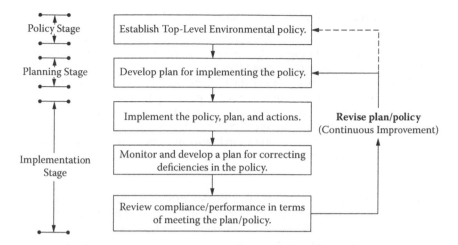

Figure 6.1 Simplified overview of a typical ISO 14001 environmental management system.

national environmental laws and regulations regarding specific environmental performance standards that must be met. In fact, it provides procedures to help ensure that all applicable laws and regulations have been identified and provides an auditing and monitoring procedure to identify any noncompliances.

The environmental management system

An ISO 14001 EMS has the standard functions described in the following subsections. Figure 6.1 depicts a simplified flowchart of the EMS process. An EMS provides a structured system (i.e., plan, do, check, revise) in which a set of management procedures is used to systematically identify, evaluate, manage, and address environmental issues and requirements. It also provides procedures and mechanisms that help ensure that necessary actions are taken to integrate environmental safeguards and compliance into day-to-day operations and long-term organizational planning (e.g., including government facilities and projects) which govern the organization's activities, products, and services. Some of the key features and characteristics of the EMS are described below.

Table 6.2 provides a brief overview of this process.

Essential EMS functions

The following subsections provide a more detailed description of essential EMS functions.

Table 6.2 Typical Seven-Stage Cycle ISO Development
and Maintenance

Stage 1—Environmental policy: The EMS process begins with the preparation
and establishment of an environmental policy.

Stage 2—Planning: The next stage of the EMS process involves developing a
plan for implementing the system. While the planning function is often
performed to determine how an organization will meet its quality policy, it can
also be used more comprehensively to develop detailed environmental plans.
Environmental aspects are identified, environmental objectives and targets are
established, and a program to achieve them is developed. This plan includes
identification of:

1. Environmental Aspects: Operations, activities, products, and services are
   reviewed to identify how they interact with and may affect the environment.
2. Legal and Other Requirements: The plan identifies legal and other
   requirements that apply to the organization's environmental aspects.
3. Objectives and Targets: Environmental objectives and targets are developed
   and communicated throughout the organization. A program is developed
   for achieving objectives and targets.

Stage 3—Implementation: Once the plan has been formalized, the EMS is ready
for actual integration and implementation with the organization's functions
and activities. Environmental and EMS responsibilities are assigned. Employees
are trained to ensure that they are aware of the plan and are able to perform
required duties in compliance with the EMS policy and plan. Specific work
procedures are developed, defining how specific tasks are to be conduced.
These implementation requirements are summarized below:

1. Structure and Responsibility:
   a. Roles, responsibilities, and authorities are defined for personnel whose
      activities may directly or indirectly affect the environment.
   b. Individuals are apppointed by top management as the management
      representatives. They are assigned responsibility and authority for
      ensuring that the EMS complies with the ISO 14001 standards and for
      reporting EMS performance to top management.
2. Training, Awareness, and Competence:
   a. The organization identifies training requirements of personnel whose
      work may significantly impact the environment. The personnel must
      receive appropriate education and training, and/or have experience to
      deal with environmental requirements.
   b. Communication: Communication of relevant information concerning
      environmental aspects is required throughout the organization.
   c. Environmental Management System Documentation: Information must be
      maintained, which describes the basics of the EMS. The documents must be
      reviewed on a regular basis. This documentation must be managed and
      maintained through an established document control system (DCS).

Table 6.2 Typical Seven-Stage Cycle ISO Development
and Maintenance   (Continued)

d. Operational Control: Activities that can significantly impact the environment, and are relevant to the organization's objectives and targets, must be identified. The organization must ensure that these operations are performed according to the EMS plan to ensure they are performed under *controlled conditions*. Controlled conditions can include documented procedures with specific operating criteria.

e. Emergency Preparedness and Response: The organization must identify potential accidents and emergency situations that may result in an environmental impact. Procedures must be developed for responding to such accidents and emergency.

Stage 4—Monitoring and corrective action: This stage involves checking and audits, control of nonconformances, corrective action, and preventive action. Characteristics of operations and activities that can significantly impact the environment need to be regularly monitored and measured. Monitoring and measurement results need to be compared with legal and other requirements to assess compliance.

Stage 5—Management review: The final stage involves a review by the organization's management of the EMS. This step helps ensure that the system is operating effectively and provides the opportunity to address changes that may be made to the EMS.

Environmental policy

The ISO 14001 standard requires the establishment of a high-level organizational environmental policy statement from top management that establishes an environmental commitment and direction for the entire organization. The policy is important because it provides the programmatic direction and goals of the organization (business, company, government agency, etc.). It also provides direction for the EMS process.

The policy is unique and should be tailored to each organization. It must be appropriate for the size and complexity of the organization, but generally should not exceed one or two pages in length. The policy must also be written in nontechnical language so that it can be understood by a typical reader. It is communicated to all employees and must also be made available to the public.

The policy must include senior management's commitment to (1) pollution prevention, (2) continuous improvement throughout the organization, and (3) compliance with applicable environmental regulations and standards that affect the organization. The policy provides a starting point for establishing the organization's EMS objectives and targets (described below).

Planning function

The planning function is often performed to determine how an organization will meet its quality policy. However, it can also be used more comprehensively to develop detailed environmental plans. The EMS team identifies legal requirements and also considers how the organization's functions and activities interact with the environment. The team develops a plan for reducing adverse environmental aspects of its operation. The plan spells out the details of how the EMS will be implemented. The planning stage should include employees from all levels within the organization.

Emergency Preparedness and Response   Procedures must be developed, communicated, and tested to help ensure that any unexpected incidents are effectively and efficiently responded to by internal and external personnel. A process must be established for identifying the potential emergencies, as well as procedures for mitigating their effects.

Environmental aspects

One of the more important functions that an EMS can contribute involves identifying an organization's significant *environmental aspects*. The term is defined broadly to mean the organization's activities, products, or services that interact with the environment. In contrast, an *environmental impact* is how a given aspect actually affects or changes the environment.

A systematic and verifiable procedure must be followed to identify environmental aspects and determine which ones may significantly affect environmental quality. ISO 14001 does not describe what aspects are significant, nor does it specify how *significance* is to be determined (more about this in "The Complementary Benefits of Integrating a Consolidated EIA/EMS with the Goal of Sustainable Development"). Provisions in ISO 14001, 14004, and 14031 help in identifying significant environmental aspects. The EMS policy and plans for improving environmental performance are documented and communicated to the employees.

Objectives and targets

After the environmental aspects have been identified, attention turns to developing a plan for reducing them. *Environmental objectives* and *targets* are established to meet the goals documented in the organization's *environmental policy*. "Objectives" refers to general long-term goals. In contrast, "targets" are more specific, measurable events. Targets will generally vary over time and across organizational functions and activities.

Objectives and targets need to be defined for appropriate functions and levels of the organization, and should be measurable, where practicable. The EMS is designed to identify specific objectives or targets, and describes the means to achieve them. The EMS process must be designed to ensure that objectives and targets are consistent with the environmental policy, including a commitment to compliance with legal and other requirements, continual improvement, and pollution prevention. Work procedures, instructions, and controls are developed to ensure that implementation of the policy and that the targets can be achieved.

Defining Objectives   The organization identifies its principal environmental objectives. As indicated above, environmental objectives tend to be overriding considerations, such as the development of better employee environmental training or improved environmental communication with other interested parties. These objectives will become the primary areas of focus within the improvement process.

Defining Targets   In contrast to objectives, environmental targets tend to be specific points or items such as the reduction of energy utilization by 20% or a reduction in sulfuric acid waste by-products by 10%. Targets should quantify the organization's commitment to an environmental improvement. Table 6.3 shows the difference between objectives and targets, as well the assignment of responsibilities for ensuring that these objectives and targets are met.

Identifying legal and other requirements
An EMS procedure must be established to ensure that applicable legal and other requirements are identified, and also ensure that this information is relayed to key organizational functions.

Table 6.3 Environmental Objectives and Targets for a Small Company

Objectives	Targets	Responsibility
1. Comply with all applicable environmental laws and regulations	Zero penalties or fines per year	Principal regulatory manager
2. Minimize waste and prevent pollution	Recycle 75% of all paper products, and 50% of aluminum waste	Chief process engineer
3. Energy conservation	Reduce electricity consumption by 20%	Chief plant engineer
4. Improve the EMS	Obtain ISO 14001 certification	EMS program manager

EMS documents and document control

The ISO 14001 standard requires organizations to establish procedures for controlling documents related to the implementation of the EMS. This documentation must provide guidance for effectively planning, implementing, and controlling processes, and must be sufficient to demonstrate conformance with the ISO 14001 standard. Specific procedures must establish a document control system to ensure that documents demonstrate that the organization's operations conform to the ISO 14001 standard. These procedures must ensure that documents are approved prior to use, and are reviewed and updated as necessary. Procedures must ensure that such records are identifiable, retrievable, secure, and traceable. The EMS could also be used to maintain sustainability plans and NEPA documents.

Monitoring and measurement

To gauge progress, an EMS must be monitored to determine its effectiveness, and to provide data for improvement. Many environmental impact assessment (EIA) processes such as the US National Environmental Policy Act (NEPA) do not have enforceable requirements to perform post-monitoring. Thus, the NEPA lacks a systematic process for ensuring that the final decision, including any adopted mitigation measures, is properly implemented. In contrast, monitoring is a basic element inherent in an EMS. A properly integrated EIA/EMS system can contribute to environmental protection, as it ensures that monitoring procedures are likewise executed (see "The Complementary Benefits of Integrating a Consolidated EIA/EMS with the Goal of Sustainable Development"). These ISO 14001 requirements can also help ensure that staff in charge of maintaining sustainability measures are properly trained and that sustainability commitments are monitored.

Internal Audit   A procedure must be developed for periodically monitoring conformance with the ISO 14001 standard and EMS plan, and to assess how well the organization is managing its environmental functions and operations, including compliance with applicable environmental requirements. It can also be used to assess the performance of the EMS in terms of achieving environmental objectives and targets. The EMS procedures also must specify who is responsible for performing the audit and the means of reporting the results.

Implementing mitigation and other EIA-related commitments can be included as part of the EMS audit function (see "The Complementary Benefits of Integrating a Consolidated EIA/EMS with the Goal of Sustainable Development"). Thus, the EMS audit provides another mechanism for ensuring that an agency's EIA commitments are appropriately implemented. Similarly, these audits can help ensure that sustainability commitments are performed correctly.

Continuous improvement

The ISO continual improvement process is based on (1) monitoring, (2) nonconformance and corrective and preventive actions, and (3) management review with a commitment to improve the EMS process. It can also help ensure that sustainable management practices are improved over time.

Management Review   The organization's top management is required to periodically review the EMS to ensure it is operating as planned, and is effectively performing its intended goals and objectives. The management review provides the ideal forum for determining how to improve environmental practices in the future.

Nonconformance, Corrective and Preventive Actions   An independent internal or third-party audit monitors conformance with EMS requirements, as well as applicable environmental regulations and requirements. A *corrective action* procedure is implemented when there is an environmental incident or a nonconformity; for instance, a nonconformance may include a breach of an EMS procedure or a violation of an applicable environmental regulation. The benefit of this step is that under ISO 14001, corrective actions can be viewed as positive—unlike traditional noncompliance situations; in other words, an EMS can actually be used to put a positive "spin" on a less-than-optimal result.

As applicable, a root-cause analysis may be conducted to determine the underlying cause of an incident or noncompliance; corrective actions are then taken to ensure that the problem does not happen again.[1] Findings or recommendations resulting from EMS monitoring and auditing phase provide the basis for identifying and managing preventive or corrective actions. Such preventative or corrective actions can also be designed to promote and maintain sustainability goals and objectives (see "The Complementary Benefits of Integrating a Consolidated EIA/EMS with the Goal of Sustainable Development").

A *preventative action* program is performed in the same manner as corrective actions, except that there will be no actual incident or nonconformity to address. Instead, emphasis is placed on identifying potential future problems, and taking measures to prevent them before they occur. Managers sometimes encounter opposition attempting to justify a preventative action program (when no actual incident or nonconformance has actually occurred) because it is often difficult to determine the effectiveness of a resulting initiative that prevents problems from occurring.

Implementation requirements

The following subsections outline key requirements for effectively implementing the EMS.

Responsibilities

Top management is responsible for ensuring that resources are available so that the EMS can be effectively implemented. EMS roles, responsibilities, and authorities must be defined and communicated.

Competence, training, and awareness

Persons performing tasks that have or can result in a significant environmental aspects or relate to the legal and other compliance requirements must receive appropriate training to ensure that they are competent to perform their tasks. An EMS procedure must ensure that such persons are aware of the need to comply with all EMS requirements and what they specifically must do. This requirement can help ensure that the EIA decision is correctly, effectively, and safely implemented.

Communications

The EMS must include internal and external communication procedures. ISO 14001 only requires that procedures be established, and allows an organization to decide the degree of openness and disclosure of information to the public.

Operational control

Operational controls must be established to ensure that critical functions related to the policy, significant aspects, objectives and targets, as well as legal and other requirements are properly identified.

The complementary benefits of integrating sustainability with a consolidated EIA/EMS

This section builds on the EMS concepts presented in "The Environmental Management System" by describing the complementary benefits that exist between an EMS, EIA, and the global environmental policy goal of sustainable development. This section provides the basis for the *integrated EMS/EIA/sustainable development process* described in "Constructing an Integrated EIA/EMS/Sustainable Development Process." The intent of this section is not to repeat the EMS concepts presented in "The Environmental Management System," but instead to emphasize the similarities, differences, and general complementary nature between an EMS and EIA process. EIA, EMS, and the goal of sustainable development provide three separate and independent approaches for protecting the environment. An EIA process such as NEPA provides a scientifically based process for rigorously and objectively evaluating alternatives to a proposal/plan. In contrast, an EMS provides an ideal system for implementing and monitoring an agency's EIA plan and final decision. A detailed assessment of these two processes demonstrates that both systems share many common features, and that the weaknesses of one process frequently tends to be counter-balanced by the strengths of the other. Properly combined, an integrated EIA/EMS provides an efficient mechanism for evaluating and implementing agency actions. An approach for integrating these two processes has been published in the companion text entitled *NEPA and Environmental Planning*.[2]

Meanwhile, the goal of sustainable development is to promote systems and practices. Yet, this goal typically lacks a general-purpose system for identifying, evaluating, and implementing a sustainable development plan. This section further expands upon earlier systems developed by the author by describing a fourth-generation system that uses an integrated EIA/EMS system to develop and implement a sustainable plan/program.* The advantage of this consolidated process is that it draws from the synergistic strengths of an integrated EIA/EMS to identify, plan, evaluate, and implement sustainable measures for proposed plans, projects, or programs.

Historical development of the integrated EIA/EMS system

At the request of Dr. James Roberts (former president of the National Association of Environmental Professionals [NAEP]), the author was asked to prepare a report in 1997 investigating the commonalities, strengths, and weaknesses between an integrated National Environmental Policy Act (NEPA) and an International Organization for Standardization (ISO) 14001 consistent environmental management system (EMS). This effort was in support of the US Council on Environmental Quality's (CEQ) *Improving NEPA Effectiveness Initiative.* A final report discussing the synergistic relationship between an integrated NEPA/EMS process was issued to the NAEP in 1997.[†] The report proposed a system for integrating NEPA with an ISO 14001 EMS, including a detailed analysis of the complementary relationship between NEPA and an EMS. This approach was reviewed by over a dozen NEPA and ISO 14001 specialists from the United States, and was presented

---

* This section is based on Integrating sustainable development with a consolidated NEPA/ ISO 14001 EMS redux, *Journal of Environmental Practice*, 12(1), 2010.
† The original report, "A Conceptual Strategy for Integrating NEPA with an Environmental Management System," was prepared by the author in 1997 and issued by the President of the National Association of Environmental Professionals to CEQ in 1999/2000.

Table 6.4 Key Advantages of an Integrated EIA/EMS

Establishes a policy and plan that meets the expectations of both an EIA and an EMS

Enhances planning, consideration, and analysis of environmental aspects that can adversely impact the environment

Includes robust EIA procedures with identifying the environmental aspects under an EMS

Integrates EIA documents and schedules into EMS objectives and targets

Incorporates the EIA documents and administrative record into an EMS records management system

Incorporates EIA mitigation commitments with related regulatory requirements and EMS objectives and targets

Monitors the selected EIA alternative (e.g., course of action) and implementation of applicable mitigation and monitoring commitments

at national environmental conferences. Beginning in 2000, the author began teaching workshops at NAEP conferences that outlined this process. Comments received from these workshops were incorporated to continually improve the integrated system.

The integrated NEPA/ISO 14001 approach was published in the first of several papers beginning in 1998.* With the assistance of Ron Deverman (current president of the NAEP), this report was reviewed and approved by the NAEP board of directors in 2000. The NAEP president issued the final report to the CEQ with a recommendation that it be promoted to all federal agencies. The reader is referred to the author's text, *NEPA and Environmental Planning,* for a detailed explanation of how NEPA and an ISO 14001 EMS can be effectively integrated into a single complementary system.[3]

In 2002, this approach was generalized to describe a process for integrating any EIA process with an ISO 14001-consistent EMS.[†] Later still, this approach was expanded to incorporate adaptive management (AM).[‡] Eventually, the CEQ issued guidance for integrating NEPA with an ISO 14000 EMS.

Some of the key advantages of an integrated NEPA/EMS system are depicted in Table 6.4.

---

* The original report to the NAEP was published in the *Environmental Quality Management Journal,* A Strategy for Integrating NEPA with an EMS and ISO-14000, John Wiley & Sons Inc., Spring 1998.

† Eccleston, C. and Smythe, R. Integrating Environmental Impact Assessment with Environmental Management Systems, *Environmental Quality Management Journal,* John Wiley & Sons Inc., 11(4), Summer 2002.

‡ See Eccleston C. H., *NEPA and Environmental Planning: Tools, Techniques, and Approaches for Practitioners,* pp. 47–52, Boca Raton, FL: CRC Press, 2008.

Table 6.5 EIA, EMS, and Sustainability Contributions to Integrated
Environmental Planning and Implementation Process

Component	Function
EIA	*Planning and Assessment Process*: An EIA process such as NEPA provides a robust, comprehensive, and general-purpose environmental planning process that can be used to evaluate the impacts and alternatives to proposed actions.
Sustainability	*Environmental Goal*: Provides an overarching and unifying environmental goal applicable to most programs and projects.
ISO 14001 (consistent) EMS	*Management System*: Provides an internationally accepted system for managing environmental policies, procedures, and requirements.

Integrating EIA, EMS, and sustainable development

The section further expands upon these earlier concepts by generalizing the process to address and incorporate *sustainable development* into a synergistic EIA/EMS process. Before describing how an EIA process such as NEPA can be integrated with an EMS and sustainability, it is instructive to explore the fundamental functions of these three environmental elements (EIA, EMS, sustainable development). As depicted in Table 6.5, each of the three components contributes a unique and essential function to the integrated system.

To simplify the approach, this section does not describe how this process can be integrated with AM. For an in-depth explanation of integrating AM with a consolidated EIA/ISO 14001 system, the reader is referred to *NEPA and Environmental Planning* for an in-depth explanation for integrating AM into an EIA/ISO 14001 system.[4]

How an EMS and EIA complement each other

Chapter 5 described the principles for an EIA process such as NEPA. Table 6.6 compares and contrasts some of the principal strengths and weaknesses of EIA, EMS, and sustainability. They all possess inherent strengths and weaknesses; moreover, a weakness in one component of an integrated system often tends to be offset by the strengths of one of others.

Some of the succinct characteristics outlined in Table 6.6 are described in the following subsections.

Table 6.6 EIA, Sustainability, and ISO 14001-Consistent EMS Complement Each Other

Characteristic	Sustainability	EIA	EMS
Goal	Achieve and maintain a sustainable system or component of a system.	The goal is to provide environment protection by ensuring that environmental impacts are considered during the early decision-making process.	The goal of an ISO 14001-consistent EMS is to provide a system for managing actions that affect the environment. Its continual improvement system can further help in reducing environmental aspects.
Environmental policy	The EIA/EMS environmental policy can be developed to incorporate elements of sustainability.	NEPA's policy goals (Section 101) provide a high-level commitment to protect the environment. For instance, the regulations state that NEPA analyses should be prepared for new federal policies that may significantly affect environment quality. Consistent with such guidance, an EIA can be prepared to develop an environmental and sustainable development policy.	In conjunction with the EIA process, the EMS must state its commitment to environmental protection and compliance.

Substantive mandate	Provides a goal or direction for achieving substantive and sustainable environmental performance.	Impacts, alternatives, and mitigation measures must be rigorously investigated to identify actions and alternatives that can protect the environment. However, most EIA processes lack a legally binding substantive mandate to choose an alternative that protects the environment.	Under an EMS, substantive actions are "expected" to lead to continual improvement in environmental performance (and thus environmental protection). Targets and objectives also provide tangible criteria for measuring the success in improving environmental performance.
Planning function	Provides a mechanism for defining and assessing the effectiveness of potential sustainable development plans.	An EIA provides a rigorous and comprehensive environmental planning process (sustainable development plan), but lacks an environmental system for ensuring that planning decisions are properly executed.	Requires a planning function, and provides a system for ensuring that the plan is appropriately implemented. ISO 14001 does not prescribe a detailed process (like that in an EIA) for performing the planning function.

(Continued)

Table 6.6 EIA, Sustainability, and ISO 14001-Consistent EMS Complement Each Other   (Continued)

Characteristic	Sustainability	EIA	EMS
Impact assessment requirements	Can be used to assess the effectiveness of a sustainability plan, and discriminate between alternative sustainability plans.	Most EIA processes specify detailed direction for performing an analysis of direct, indirect, and cumulative impacts.	An EMS must identify environmental aspects or actions that can impact the environment. However, little specificity concerns requirements for performing this investigation. Moreover, the assessment of environmental aspects is generally much less rigorous than most EIA processes such as NEPA's requirement to assess environmental effects.
Objectives and targets	Can be used to establish objectives and targets for a sustainability plan.	The EIA analysis can be used to identify, access, and choose objectives and targets.	Under an EMS, an organization is expected to adopt environmental objectives and targets to address significant aspects.
Significance	Can be used to identify, evaluate, and focus on significant sustainability issues.	Most EIA processes have detailed direction for determining the significance of an impact. For instance, in addition to context, ten specific factors are detailed in the NEPA implementing regulations for assessing significance.	Unlike most EIA processes, ISO 14001 lacks detailed direction for interpreting or determining the meaning of "significance."

External input	An EIA provides a mechanism for the public to provide input into developing a sustainability plan.	Most EIA processes have well-defined public participation procedures; they specify a detailed public participation and a formal public scoping process for identifying actions, impacts, and alternatives (sustainability plan), and for eliminating nonsignificant issues from further review.	An EMS simply requires that procedures (not necessarily public) be used to record and respond to external parties; however, ISO 14001 does not prescribe detailed requirements for accomplishing this task.
Other environmental requirements	An integrated EIA/EMS system provides a means for identifying and incorporating environmental requirements into a sustainable development plan.	Most EIA processes have extensive direction for performing the analysis. For instance, CEQ guidance and executive orders direct federal agencies to integrate pollution prevention (P2) measures, environmental justice, biodiversity, and other considerations with NEPA.	A top-level environmental policy is required, including a commitment to P2, which is very broadly defined.
Mitigation	Under most EIA processes, mitigation measures can support sustainable development measures, while an EMS provides a mechanism for implementing such measures.	Most EIA processes require that mitigation measures be identified and analyzed, but many do not require that such measures must be implemented.	An EMS provides a system that can be used to ensure that mitigation measures are properly executed.

(Continued)

Table 6.6 EIA, Sustainability, and ISO 14001-Consistent EMS Complement Each Other   (Continued)

Characteristic	Sustainability	EIA	EMS
Emergency preparedness	No built-in regulatory mechanism to ensure that a sustainable development plan addresses potential emergency situations; an integrated EIA/EMS system provides a mechanism for doing so and mitigating the risk of potential incidents consistent with a sustainability plan.	An EIA process can provide a rigorous planning mechanism for identifying potential incidents, and assessing the impacts, alternatives, and measures for mitigating potential threats.	EMS procedures must provide measures for preventing and responding to emergencies.
Non-conformity, and preventive and corrective actions	No built-in regulatory mechanism to ensure that a sustainability plan is correctly implemented; an EMS provides such a mechanism. An EMS can also include an adaptive process for improving the implementation of a sustainability plan. For instance, NEPA's concept of adaptive management (AM) provides an efficient corrective action mechanism for dealing with uncertainty or changing circumstances.	Under most EIA processes, organizations are responsible for ensuring that decisions and commitments are carried out. However, many EIA processes such as NEPA lack a rigorous system or procedure for ensuring such compliance once the EIA process has ended. However, a NEPA AM system can provide an effective management process for implementing corrective actions as a result of new information or changing circumstances.	An EMS must include procedures for identifying and correcting nonconformance. ISO 14001 specifies detailed procedures that can be used to (1) identify circumstances where EIA commitments or mitigation measures are incorrectly implemented, (2) correcting nonconformities, (3) mitigating their impacts, and (4) developing plans for avoiding nonconformities.

Records and documentation	An EMS provides a mechanism for managing a sustainable development plan and other important records.	Many nations such as the US require EIA documents to be maintained as part of the administrative record. However, most EIA processes do not specify how such a system should be maintained. An EMS can be used for maintaining EIA records.	An EMS specifies detailed procedures for controlling and maintaining records needed to demonstrate conformance with the EMS standard.
Monitoring	An EMS provides a mechanism for monitoring the progress of a sustainability plan.	Many EIA processes such as NEPA encourage (and sometimes require) post-monitoring measures. However, little direction is provided in terms of how monitoring should be performed.	Monitoring is mandated as part of the EMS continual improvement cycle. Specific direction is provided on how this element is to be performed.
Continual improvement	An EMS provides a mechanism for continually improving the implementation of the sustainability plan.	Most EIA processes provide no direction for performing a continuous improvement process. However, under NEPA, the CEQ has promoted a cyclical process known as adaptive management.	A continual improvement process is inherent in an EMS.

(Continued)

Table 6.6 EIA, Sustainability, and ISO 14001-Consistent EMS Complement Each Other   (Continued)

Characteristic	Sustainability	EIA	EMS
Audits	An EMS provides a system for auditing the success of and compliance with a sustainable development program. It also ensures that the project is implemented according to the selected course of action (EIA process) that EIA impacts remain within designated parameters, and that mitigation measures are correctly implemented to promote sustainable commitments.	Most EIA processes lack a well-defined auditing process. However, conformance and commitments may be efficiently reviewed and audited where linked to EMS objectives and targets. An EMS audit provides a means of ensuring that the EIA process and commitments are correctly implemented.	ISO 14001 defines specific internal auditing requirements for periodically assessing conformity with the EMS; the results must be presented to management for review.
Management review	ISO 14001 requires top management to review the EMS progress; this could also include progress in implementing the sustainable development plan.	Most EIA processes require the responsible decision maker to review the EIA document and choose a course of action. Beyond this direction, there is often no requirement that management periodically review the implementation of the selected course of action.	Under ISO 14001, top management is required to periodically review the progress in meeting EMS requirements.

Developing policies and plans

Many EIA processes, including NEPA, recognize four broad categories of activities (i.e., policies, programs, projects, and plans) as potentially subject to a detailed impact and alternative analysis. For instance, under NEPA, establishment of federal policies and plans are "actions" potentially subject to a full NEPA assessment. Thus, policies and plans established as part of an EMS may potentially be subject to EIA requirements, particularly where a policy or plan entails potentially significant environmental impacts or issues.

Although an environmental planning function is a mandatory element within an EMS, the ISO 14001 standard provides only limited specifications for performing the planning function. For example, specific procedures and requirements regarding scoping, investigating environmental aspects, defining temporal and spatial bounds, interpreting significance, and other requirements are only vaguely inferred or defined.

In contrast, most EIA processes provide highly prescriptive direction and requirements for ensuring that an accurate and scientifically defensible planning and analysis process is followed to provide decision makers with information sufficient in reaching an informed decision. This can also include investigating, analyzing, and comparing alternative sustainability plans. Moreover, these requirements are in many cases reinforced by decades of experience gained through engagement with diverse missions and environmental issues. Properly integrated, a combined EIA/EMS system can provide a synergistic process for planning sustainable actions and implementing decisions in a manner that protects and enhances environmental quality and sustainability, while reducing cost, generation of pollutants, and consumption of strategic resources.

Substantive versus procedural requirements

As described earlier, most EIA processes such as NEPA are not obligated to select an environmentally preferable alternative or to demonstrate that its decision conforms to the environmental goals (i.e., substantive mandate), such as those established in Section 101 of NEPA. Thus, the EIA's contribution is derived not from a *substantive* mandate to choose an environmentally beneficial or sustainable alternative, but from its *procedural* provisions, which require agencies to rigorously evaluate and seriously consider the impacts of potential actions in their final decision, just as they would balance other more traditional factors such as cost and schedules.

In contrast, an ISO 14001-consistent EMS involves a general expectation that some type of substantive action will be taken to improve environmental quality. Not only are environmentally beneficial actions *presumed* to be taken, but they are also implemented in a cycle of continual improvement in environmental management practices. To this end, an EMS could provide a mechanism for enacting some of the substantive environmental mandates that most EIA processes lack.

Similarly, most EIA processes require analysis of mitigation measures but places no substantive burden on decision makers to choose or enact such measures. ISO 14001, in contrast, requires organizations to establish *objectives* for improving environmental performance. Similarly, environmental *targets* are established for measuring and achieving those objectives.

Achieving these objectives could involve implementing actions similar to NEPA mitigation measures. Again, most EIA processes prescribe rigorous requirements for planning and investigating mitigation measures, while an EMS provides a mechanism for implementing such measures. An integrated EIA/EMS system could be used to continually improve on the implementation of an adopted sustainability plan.

Analysis requirements

Most EIA processes such as NEPA provide practitioners with highly prescriptive requirements for ensuring that an accurate and defensible analysis is performed, and provide a decision maker with information sufficient to support informed decision making. Most EIA processes are more demanding than ISO 14001, requiring not simply identification of environmental aspects, but a comprehensive analysis of the actual direct, indirect, and cumulative impacts on environmental resources.

As described earlier, NEPA (and some other EIA processes) practice is reinforced by more than four decades of experience accumulated by a diverse range of federal agencies, each with its own mission and often unique environmental issues. EIAs provide a more comprehensive and rigorous planning process than ISO 14001 for ensuring that environmental impacts are identified, evaluated, and considered before a decision is made to pursue an action. The EIA analysis can be used to evaluate and compare the advantages and disadvantages, as well as compare and discriminate between various sustainable development plans.

Assessing significance

Both EIAs and ISO 14001 specify requirements for assessing *significance*. The concept of significance permeates most EIA processes such as NEPA's regulatory provisions. For instance, the NEPA regulations include detailed definitions of "significance" and, in addition to "context," provide ten specific factors decision makers are required to consider in making such determinations (40 CFR §1508.27).

In contrast, under ISO 14001, "significance" is defined vaguely and contains little direction for determination. Again, most EIA processes bring many years of experience to bear on the problem of how best to determine "significance." This EIA analysis can be used to evaluate and discriminate in terms of significance, the advantages and disadvantages of various sustainability plans.

Public involvement

Public input, review, and participation are essential to nearly every EIA process. For instance, the NEPA public scoping process is designed to solicit comments from the public, potentially affected parties, government agencies with special expertise, and subject matter experts. The public is consulted with respect to the scope of the NEPA analysis; the federal agency must also allow the public to review and submit comments on the NEPA analysis.

In contrast, ISO 14001 provides no requirement for *public* scoping and participation, only a requirement to develop a plan (not necessarily public) for external communications and inquiries. Lack of such requirements can be viewed as a weakness in many parts of the ISO 14001 standard. Thus, the EIA experience combined with transparency, public participation, and scoping helps to balance the weaknesses of an EMS. An integrated system can facilitate the public's ability to shape and participate in the development and implementation of a sustainability program.

Incorporating pollution prevention measures

The CEQ has issued guidance indicating that, where appropriate, pollution prevention (P2) measures are to be coordinated with, and included in, the scope of a NEPA analysis.[5] Some other EIA processes also have issued similar directives. ISO 14001 speaks to the merits of P2, but primarily from the standpoint of establishing a top-level policy committed to P2. ISO 14001 provides a top-down policy for ensuring that P2 is actually incorporated at the operational level.

In comparison, most EIA processes provide an ideal framework for planning and evaluating the effectiveness of a comprehensive P2 strategy or plan. An integrated system may help facilitate the development of a sustainability plan that reduces pollutants, while encouraging recycling, and other beneficial environmental practices. *NEPA and Environmental Planning*, provides direction for combining P2 with an integrated NEPA/EMS.*

Incorporating other environmental requirements

To the extent feasible, federal agencies are instructed to integrate NEPA with other environmental reviews (e.g., regulatory requirements, permits, agreements, project planning, and policies) so that procedures run concurrently rather than consecutively; this requirement reduces

---

* Eccleston, C. H., *NEPA and Environmental Planning: Tools, Techniques, and Approaches for Practitioners*, pp. 47–52, Boca Raton, FL: CRC Press, 2008.

duplication of effort, delays in compliance, and the overall cost of environmental protection.[6] Specifically, NEPA requires federal agencies to

- *Identify* other environmental review and consultation requirements ... prepare other required analyses and studies **concurrently** with, and **integrated** with, the environmental impact statement... (§1501.7[a][6], emphasis added);
- **Integrate** the requirements of NEPA with other **planning** and environmental review procedures... (§1500.2[c], emphasis added); and
- [combine] *Any environmental document in compliance with NEPA ... with any other agency document...* (§1506.4, emphasis added).

Consistent with the NEPA's regulatory direction, ISO 14001 expects organizations to identify applicable legal and regulatory requirements. The intent is to ensure that organization activities meet applicable legal and regulatory requirements. To this end, an EMS provides a system that can identify regulatory and other requirements, and ensure that they are incorporated into an integrated EIA/EMS process. As detailed in the three aforementioned regulatory provisions, a combined EIA/EMS system can be used to develop and implement a sustainability program.

Constructing an integrated EIA/EMS/ sustainable development process

As shown in Figure 6.2, the EIA planning process is generally triggered through identification of a need for a proposed action (i.e., a new policy, program, plan, or project). The EMS is capable of managing a full range of construction and operational activities (i.e., proposals, operations, services).

To develop an efficient and effectively integrated system, EIA, EMS, and sustainability practitioners must collaborate closely. By establishing applicable environmental objectives and targets, an EMS can help ensure environmental protection and implementation of sustainability commitments and mitigation measures through a monitoring program. Conceptually, the consolidated system (see Figure 6.2) is composed of three integrated stages or phases:

Phase 1: EIA, EMS and Sustainable Development Policy
Phase 2: EIA
Phase 3: EMS

In Phase 1, the EIA and EMS processes are used to develop a high-level environmental and sustainability policy (Figure 6.2, Blocks 1 and 7). In some instances, this could be accomplished by preparing an EA that investigates and compares the impacts and effectiveness of an environmental and sustainability development policy.

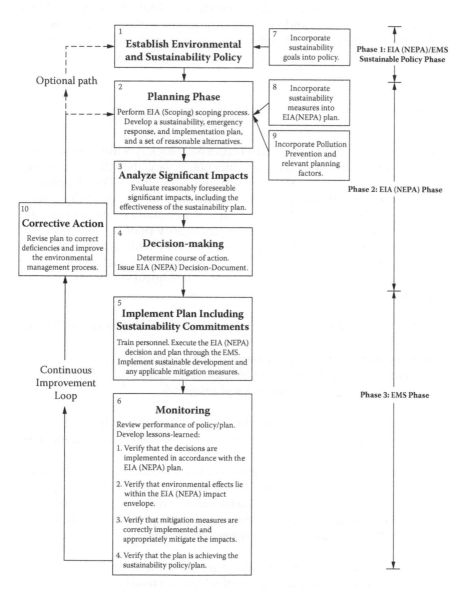

Figure 6.2 Approach for incorporating sustainability development commitments into an integrated EIA (NEPA) and EMS process.

In Phase 2, the EIA process is used to identify, plan, evaluate, and compare alternatives. This phase is also used to select a course of action, including a sustainable development plan (Figure 6.2, Blocks 2, 3, 4, 8, and 9).

In Phase 3 (Figure 6.2, Blocks 5, 6, and 10, as well as a "continuous improvement" loop), the EMS is used to implement and effectively manage the course of action selected during Phase 2. EMS elements would be used to monitor, audit, and continuously improve the course of action, including a sustainability plan that was selected in Phase 2. The principal components of this integrated system are described below.

Phase 1

Policy

The integrated system described in Figure 6.2 (Block 1) is initiated with the establishment of a high-level environmental policy, including a commitment to environmental quality and sustainable practices. The EIA process is used to scope and define the environmental and sustainability policy. The policy will also include a commitment to pollution prevention. EIA procedures will be followed in making this policy publicly available.

The sustainable development policy will vary with the context of the proposal under consideration. For example, potential sustainable development policies for a proposal that affects a biological system or a renewable resource might involve concepts such as

- Managing water resources so they do not exceed the safe (replenishable) yield
- Sustainably managing the harvest of a forest
- Maintaining and enhancing biological diversity
- Preserving and rehabilitating soil resources

The reader should note that developing a sustainable policy for a facility, city, or nonrenewable resource may require special consideration with respect to developing policies and plans.

It is important to note that the concept of sustainable practices varies with the context of resources. Sustainable practices may need to be applied differently to problems involving nonrenewable resources as compared to the way they would be applied to renewable resources. For instance, consider the following sustainable development example:

> The program will strive to increase the longevity of nonrenewable sources of silver by ensuring that minerals are mined and used in acceptable ways, both economically and environmentally.

Many sustainability specialists would question whether the afore-mentioned statement strictly meets the underlying goal of sustainability. It could be argued that mining any ore body is an unsustainable practice because mining cannot continue indefinitely and is therefore incompatible with a strict definition of sustainability. An alternative approach might involve developing a sustainable policy or plan that emphasizes substitution of renewable resources or recycling of nonrenewable resources. A second approach might involve developing a policy or plan in which mining is performed to ensure that environmental quality is sustainable over time.

Phase 2

Planning

The EIA planning phase (Figure 6.2, Block 2) essentially begins with a formal public scoping process. The EIA scoping process is used to obtain public input and separate potentially significant issues from those that are nonsignificant. The scoping process identifies the scope of actions, alternatives, potentially significant environmental aspects (which will then be mapped into environmental impacts for detailed analysis), and mitigation measures. If appropriate, this analysis can also include development of an emergency response plan (ISO 14001 requirement), thus satisfying an important EMS requirement.

With respect to sustainability, this phase is also used to develop specific measures that can be taken to support the environmental and sustainability policy (Block 1). The final result is a detailed EIA plan (description of the proposed action and alternatives) that also includes specific measures that would be taken to achieve the stated sustainability policy. Consistent with an EIA process, different alternatives are investigated. For example, under NEPA this detailed plan might involve preparing a description of the proposed action and alternatives. As applicable, the plan should also include measures for controlling or reducing pollution.

Figure 6.3 shows some of the factors (e.g., public input, technologies, cost versus benefits, etc.) for developing an integrated environmental/sustainability plan. A proposed sustainability plan should be reviewed to ensure that it is consistent with other policies, including laws and regulations. If it is not consistent, the plan should be revised to do so (see loop in Figure 6.3).

Determining the appropriate sustainability scale and context

Virtually all human activities influence sustainability to some degree. With respect to the planning function, sustainable goals can be studied

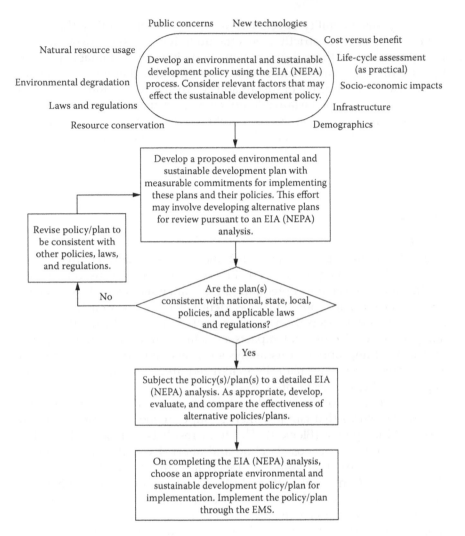

Figure 6.3 Integrating a sustainability development plan with the EIA (NEPA) and EMS process.

over an array of different time periods, contexts, and scales (levels or frames of reference) of environmental, economic, and social organization.[7] This focus can range from the global context down to the sustainability of a national program, city, community, facility, or an individual project. The context can also range from settings as varied as entire ecosystems, forests, oceans, or agricultural areas. During the EIA scoping process, it is essential that the planning team choose the appropriate context and scale for the sustainability study.

Measuring Sustainability EIA analysis should define criteria for instituting sustainability practices. Environmental objectives are documented in the sustainability plan. For instance, depending on the scope and nature of the action (policy, program, plan, or project), potential environmental objectives might include the preservation or maintenance of one or more of the following resource elements:

- Biota (e.g., minimizing impacts, achieving a carrying capacity, promoting sustainable yields)
- Land usage (e.g., achieving sustainable development and minimizing urban sprawl)
- Energy (e.g., conservation of nonrenewable and renewable energy sources)
- Natural resources (e.g., conservation of nonrenewable and renewable natural sources)
- Water (e.g., conservation and recycling programs)
- Materials (e.g., conservation and recycling of nonrenewable, and renewable materials)
- Food (e.g., sustainable agricultural practices)
- Waste (e.g., recycling, pollution prevention, waste minimization)

A large number of sustainability indicators, metrics, indices, and benchmarks have been established.[8] These include environmental, economic, and social measures (separately and integrated) over many contexts, as well as spatial and temporal scales. Measurable environmental targets would then be defined to monitor progress in achieving these objectives.

Analysis, significance, and decision making

The proposed plan is then subject to an EIA analysis that evaluates the impacts and effectiveness and various alternatives, including sustainable development options. For instance, under NEPA, an EA or EIS would be prepared to assess the potential impacts and compare the alternatives to the proposal (Figure 6.2, Block 3).

The analysis also investigates the beneficial and adverse impacts associated with the sustainability plan, as well as the probable effectiveness of these planned activities. Such impacts should be investigated in terms of the three pillars of sustainability: environmental, economic, and social factors. As described earlier, most EIA processes such as NEPA define significance. For instance, in addition to the environmental context, NEPA provides ten significance factors (40 CFR §1508.27[b]) for assessing significance and reaching a final decision.

The agency's responsible decision maker then reviews the proposal (including the sustainability plan) and forecasted impacts, and reaches a final decision regarding the course of action, including a sustainable development plan (Figure 6.2, Block 4).

Phase 3

Implementation

After a final decision has been reached using the EIA process, the course of action including mitigation and sustainable development measures can be implemented through the EMS (Figure 6.2, Block 5). ISO 14001 ensures that key personnel (with environmental responsibilities) receive and possess job-appropriate experience or training.

Monitoring, enforcement, and corrective action phase

Depending on the size, complexity, and scope of the proposal, a centralized oversight office might be assigned responsibility for implementing the proposal and monitoring compliance (Block 6, Figure 6.2). As appropriate, an environmental compliance officer (or equivalent) could be assigned responsibility for preparing and transmitting input and status reports to the oversight office.

Monitoring data are evaluated to verify compliance with established policies, plans, and the agency's decision. As appropriate, the organizational policy or plan is revised to correct deficiencies (See "continuous improvement loop" branching to box labeled "corrective action," Block 10, Figure 6.2). As dictated by the results of the EMS monitoring and auditing steps, changes may need to be made to the methods used for managing and implementing the EIA decision and sustainable commitments.

Ultimately, the ISO 14001 expectation is that environmental aspects (impacts) will tend to dissipate with time, such that subsequent revised plans might address issues different from those in the existing plan. Such a process ensures a continuous improvement cycle, which is the hallmark of an EMS, and also promotes the EIA policy goals and paradigms as well as the goal of sustainable development.

Summary

In summary, an integrated EIA/EMS system provides an ideal system for scoping, evaluating, and developing a sustainable plan or program. Once a sustainable plan or program has been developed as part of the EIA process, the EMS element provides a particularly effective mechanism for implementing the agency's plan or program, and ensuring that it meets the sustainable development criteria evaluated in the EIA plan. A properly integrated system can provide agencies with a synergistic process for protecting and preserving environmental quality.

Endnotes

1. Pojasek, R. B., Introducing ISO 14001 III, *Environmental Quality Management* (DOI: 10.1002tqem.20154), Autumn 2007.
2. Eccleston, C. H., *NEPA and Environmental Planning: Tools, Techniques, and Approaches for Practitioners*, Boca Raton, FL: CRC Press, pp. 38–47, 2008.
3. Eccleston, C. H., *NEPA and Environmental Planning: Tools, Techniques, and Approaches for Practitioners*, Boca Raton, FL: CRC Press, pp. 38–47, 2008.
4. Eccleston, C. H., *NEPA and Environmental Planning: Tools, Techniques, and Approaches for Practitioners*, Boca Raton, FL: CRC Press, pp. 47–52, 2008.
5. Council on Environmental Quality, Guidance on Pollution Prevention and the National Environmental Policy Act, 58 FR 6478, January 29, 1993.
6. Council on Environmental Quality, Regulations for Implementing the Procedural Provisions of the National Environmental Policy Act, 40 CFR, Pts. 1500–1508, 1978.
7. Millennium Ecosystem Assessment Board, Dealing with scale. *Ecosystems and Human Well-Being: A Framework for Assessment,* London: Island Press, 2003.
8. Hak, T., et al., *Sustainability Indicators*, SCOPE 67. London: Island Press, 2007.

Acronym List

Parentheses indicate relevant chapter numbers.

BPP: best professional practice (Introduction)
CAFÉ: US corporate average fuel economy (CAFE) (2)
CATX: categorical exclusion (CATX) (1)
CBD: Center for Biological Diversity (CBD) (2)
CEARC: Canadian Environmental Assessment Research Council (1)
CEQ: US Council on Environmental Quality (1, 2)
CIA: cumulative impact assessment (1, 2)
CRU: East Anglia University Climate Research Unit (2)
DOE: US Department of Energy (2)
DOT: US Department of Transportation (2)
EA: environmental assessment (1, 2)
EIA: environmental impact assessment (Introduction)
EIS: environmental impact statement (EIS) (1, 2)
E.O.: Executive Order (2)
FAA: Federal Aviation Administration (1)
FONSI: findings of no significant impact (1, 2)
GHG: greenhouse gas (2)
GIS: geographic information system (1)
IPCC: Intergovernmental Panel on Climate Change (2)
MCL: maximum concentration level (1)
NAAQS: national ambient air quality standards (1)
NAS: National Academy of Sciences (3)
NEPA: National Environmental Policy Act (Introduction)
NHTSA: National Highway Traffic Safety Administration (NHTSA) (2)
OMB: Office of Management and Budget (OMB) (2)
PEA: programmatic environmental assessment (formerly P-EA) (1)
PEIS: programmatic environmental impact statement (formerly P-EIS) (1)
PSD: prevention of significant deterioration (1)

RUS: USDA rural utility service (2)
SDP: significant departure principle (1)
SIA: social impact assessment (SIA) or socioeconomic impact
          assessment (4)
SPM: summary for policymakers (SPM) (2)
USACE: US Army Corps of Engineers (1)
VEC: valued environmental or ecosystem component (VEC) (1)

Index